HOW TO BUILD A DRAGON
OR DIE TRYING
A Satirical Look at Cutting-Edge Science

HOW TO BUILD A DRAGON OR DIE TRYING
A Satirical Look at Cutting-Edge Science

Paul Knoepfler
University of California, Davis, USA

Julie Knoepfler
Davis High School, California, USA

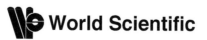 **World Scientific**

NEW JERSEY · LONDON · SINGAPORE · BEIJING · SHANGHAI · HONG KONG · TAIPEI · CHENNAI · TOKYO

Published by

World Scientific Publishing Co. Pte. Ltd.

5 Toh Tuck Link, Singapore 596224

USA office: 27 Warren Street, Suite 401-402, Hackensack, NJ 07601

UK office: 57 Shelton Street, Covent Garden, London WC2H 9HE

Library of Congress Cataloging-in-Publication Data

Names: Knoepfler, Paul, 1967–   author. | Knoepfler, Julie, author.
Title: How to build a dragon or die trying : a satirical look at cutting-edge science /
    Paul Knoepfler (University of California, Davis, USA),
    Julie Knoepfler (Davis High School, California, USA).
Description: New Jersey : World Scientific, 2019. | Includes index.
Identifiers: LCCN 2019015578 | ISBN 9789813274464 (hardcover : alk. paper)
Subjects: LCSH: Genetics--Popular works. | Genetic engineering--Popular works. |
    Science--Popular works.
Classification: LCC QH437 .K66 2019 | DDC 572.8/636--dc23
LC record available at https://lccn.loc.gov/2019015578

British Library Cataloguing-in-Publication Data
A catalogue record for this book is available from the British Library.

For any available supplementary material, please visit
https://www.worldscientific.com/worldscibooks/10.1142/11104#t=suppl

Typeset by Stallion Press
Email: enquiries@stallionpress.com

Preface

We had fun writing a book about how to build a dragon, but there was a lot more to it than we thought at the beginning. Turning an idea into a book took a couple of years.

The "how to build a dragon" idea first came from a school project that Julie did about how one might use new technologies to try to make a real dragon. I (Julie speaking here) decided to do my eighth-grade science fair project on a less than conventional subject: *How to Build a Dragon: For Fun or World Domination*. Other kids in my class did experiments such as "can I make a volcano with just clay, baking soda/vinegar, and sheer willpower?" and "Coke *versus* Pepsi: who will prevail?" Unlike my project, the other kids' experiments were actually doable, and potentially, far less dangerous, but I was up for a challenge.

Although my "theoretical" experiment was a little unorthodox, my science teacher was cool about it. I wrote a short "instruction manual" (with the help of my Dad) and built a model. The day of the class presentations finally came, and I didn't really know what to say. "Hey, class...um so, I figured out how to build a dragon." The class was actually a little stunned after my presentation by the weirdness of my "not a project" project. It turned out alright in the end though because my teacher had an open mind.

After Julie's science project, we began talking through the types of science and technology that might go into a dragon-building project. Paul's own research lab at UC Davis School of Medicine uses some of these technologies including stem cells and CRISPR gene-editing, but not to build a dragon, of course. The Knoepfler Lab's goals are to make stem cell therapies safer and to make new, better cancer treatments, particularly for the tumors that arise in kids.

At some point, whether it was on a walk or while making dinner, the following question came up between the two of us: "What if it weren't just a school assignment?" I (Paul) suggested the idea of writing a whole book together about building a dragon.

The idea of trying to build a real dragon is so over the top that right away we felt the book in a sense would be a satire of the way that real science becomes hyped so often. For these reasons, it is very important as you read to keep in mind that some of the more extreme ideas and statements in this book are intentionally that way. They are meant as a satire of science itself and of the associated hype. We hope that this book steers people away from hyping science.

Still, it's probably inevitable that at least a few people will take things too seriously and accuse us of irresponsibly promoting real-life dragon building or other crazy things. "How could you?" they'll yell, tweet, etc. They might even say we are hyping science even as we are parodying science hype. We've prepared ourselves for this and will calmly remind them of the subtitle of the book. Also, we'll encourage them to read this Preface.

It's a good thing we have thick (dragon-like?) skin.

We realized during the brainstorming stage, before we even started writing, that someone might actually try to build a dragon in the future, and our book could unintentionally help them in what would almost certainly be a disastrous project.

Hopefully, that won't happen.

As we talked and then started writing, it also became clear that the idea of making a real dragon would present many hypothetical ethical dilemmas. Dragons are monsters after all. These possible dilemmas started piling up and became an entire chapter of the book. Please take that last chapter seriously. We want to thank Paul's colleague at UC Davis School of Medicine, Dr. Mark Yarborough, who is a respected bioethicist, for his advice and feedback on that chapter.

As we did research and started writing more, it turned out that a few other people had already given a little thought to how one might try to build a dragon or at least a dragon-like creature. We want to give a big "thank you" to those who wrote articles that inspired and informed us in ways that were very helpful in writing this book. We have cited them in specific places.

We also want to thank our scientific editor, Jane Alfred, who was immensely helpful in making this a far better book. A big thanks as well goes to our editor at World Scientific Publishing Company, Yugarani. Our thanks also go to Anca Knoepfler and Dan Knoepfler for reading and providing feedback on the draft of this book.

In a sense, this book is also meant as a kind of wake-up call as well. Although we haven't seen any dragon-building projects get launched, many people are talking about making entirely new organisms with CRISPR gene-editing and other technologies.

So-called "biohackers" are discussing this kind of stuff and even sometimes trying to make it happen. Now. Something like a dragon isn't out of the question in coming decades. The same is true of other previously mythical creatures such as unicorns (see Chapter 7), which in fact would be far easier to make than a dragon.

Even if no one makes a full-blown real dragon and maybe no one will even try, they might change the world in positive or more likely negative ways by making other new creatures. A giant, glow-in-the-dark dragonfly that eats birds? A new kind of ravenous fish that lives on land? A frog that has such huge leg muscles it can jump 20 meters? Just use your imagination.

We hope you will have fun and learn many new things as you read *How to Build A Dragon*. It should also spark your imagination about science. We hope that some kids might become scientists because of this book.

Contents

Chapter 1

So, you want a dragon?

Introduction

For thousands of years, people have been fascinated with dragons. Who wouldn't want to even just catch a glimpse of one? The more audacious idea of actually owning one's own dragon has captivated people even more. Have you ever imagined owning one? We have. However, it is generally thought to be impossible to have a dragon. In this book, we challenge that assumption and explain how we'd go about making our own dragon.

But first, why do we have to go to all the trouble of making our own dragon?

Can't we just find one?

Perhaps buy a dragon egg on eBay?

Or maybe, wait until someone else makes one and see if they'll give it to us?

No. Unfortunately, none of these options are going to work.

But because we want a dragon, we're going to have to build it ourselves. It might require a lot of work, but it could also be a once-in-a-lifetime adventure. It has been fun for us to even just think about how we might make a dragon using interesting combinations cool, cutting-edge technologies. So, we are excited about dragon building even if it's potentially difficult and highly dangerous.

To build a dragon, we also needed to learn loads of new information about a wide range of existing real creatures, which although not dragons themselves, in some ways possess equally amazing powers.

Take bombardier beetles. They send potentially deadly, nearly boiling temperature explosions out of their butts as a defense mechanism, which got us thinking about how we could create a fire-breathing dragon.

Then there are electric eels, which have cool, specialized cells called electrocytes that produce electricity. Using their home-brewed electricity, eels can shock other creatures. They can also use these biological electrical systems to sense their environments like a kind of electric radar. These eels got us thinking about how our dragon might initially spark its fire – not just by using a flame, but electricity as well.

Also, just the fact that insects, birds, and bats, along with other creatures, can fly is pretty amazing. In order to fly, we humans have to resort to "cheating" – we have to use technological innovations, like planes or jetpacks. Even more remarkable is that certain enormous creatures that once roamed the planet but are now extinct – like Pteranodons – could fly. Pteranodons were about as huge as we imagine dragons could be and scientists think they looked like dragons too.

From learning all this new information, we realized that animals alive today have a whole range of "technologies" that were created by evolution, which are already surprisingly powerful and so could help us make a mighty dragon.

One of the biggest recent technological innovations goes by the name "CRISPR-Cas9 gene-editing" and came from one of the tiniest living things, bacteria. Certain bacteria use CRISPR, which is short for "clustered regularly-interspaced short palindromic repeats" (now you know why everyone uses the acronym CRISPR), as a sort of immune system to fight off viral infections.

While bacteria utilize CRISPR systems to chop up the DNA of invading viruses, researchers adapted clever CRISPR systems to be used instead to make precise mutations in the genomes of cells and even whole organisms. If you remember your biology, DNA consists of four units, called bases: A for adenine, C for cytosine, G for guanine, and T for thymine. CRISPR can be used to make as small a change as, for example, a C to T in the DNA code of the cells of just about any living thing. Alternatively, CRISPR can alter a much larger region, perhaps spanning hundreds or thousands of bases, to tailor gene function.

At the same time as we were "oohing and aahing" about the cool science already in nature, it also became clear to us – while making our dragon-building plan and writing this book – that things could go disastrously wrong for us. In fact, there are many ways we could end up dying along the way! As a result, as we explain to you our dragon-making plan, we will share with you both the cool side of our efforts but also the numerous ways things could go disastrously and even fatally wrong at every step along the way.

It's a funny, and sometimes sobering, exercise to imagine oneself dying in all kinds of strange ways. We figure that the most likely way we'd meet our tragic end would be because our dragon either incinerated us or dropped us from a high altitude while learning to fly. Or perhaps both, if it just gets annoyed with us. Imagine it dropping us from up in the sky and then swooping down to flame broil us in mid-air. A wonderful thought, right?

Glitches at any point in our plan could often lead to our demise in other mostly awful, but sometimes funny ways. Imagine being farted to death by our dragon or burped to death if our dragon cannot light the gases it makes to breathe fire. While writing, we tried to keep the possibility of death in mind despite how amazing a real dragon would be. We also had to keep our sense of humor.

And, yes, we realize that a huge disaster on the scale of *Jurassic Park* could happen if we set about building a dragon based on our plans in this book. We admit that our efforts might affect many other people in the world and not always positively. This hypothetical, large-scale risk is more likely to lead to a real disaster if we decide to make a breeding pair of dragons. They could, after all, turn out to be excellent parents, and start cranking out baby dragons. On the other hand, from a positive perspective, breeding dragons is the best way to sustain and expand our "invention." We've decided that we want to go for it, despite knowing the potential risks to us and the world.

So… where do we start?

The dragon or the egg?

To build our own real dragon, do we begin with the dragon or with the egg?

In truth, we haven't seen many, reliable reports of dragon sightings lately. So, catching a living, breathing dragon is not going to be a great plan. And even if there were dragons out there for us to try to catch, it'd be nearly impossible to nab one without it or us dying in the process. Besides, even if you could pull it off, the captured dragon could then see us as its archnemesis. Who wants to be the number one enemy of a dragon? Not us.

As for dragon eggs, they are also pretty hard to come by. Unlike in the fantasy world of *Game of Thrones (GoT)* where character Daenerys Targaryen gets a wedding gift of three eggs that turn out to be real dragon eggs, no one is going to give you a dragon egg as a present or leave one on the side of the road to hatch. Although, when researching this book, we momentarily got excited when we saw this old news article, announcing "Huge haul of rare pterosaur eggs excites paleontologists" in the journal *Nature* [1]. But, of course, the eggs from these flying reptiles were sadly just fossils.

Can you blame us for envisioning cartons of fresh pterosaur eggs? Almost as easy to find as chicken eggs in the grocery store? If only we could just pop them in an incubator, establish a breeding pterosaur colony, and then try to give them the ability to breathe fire by using exciting new gene-editing technologies, like CRISPR. We'd have made something very close to a dragon.

No one else seems to be building a dragon either that we could try to buy, at least not publicly. And dragon building technology is likely to be super expensive to invent anyway (as well as being difficult to steal). By the way, since we're speaking of unethical things like stealing…You should know that in the last chapter of this book – on the challenges and ethical dilemmas that arise in this project – we also discuss how not to turn humanity against our dragon-building effort.

And there in Chapter 8 we also present some ideas of how to get the big money we will need for research honestly and without selling out to dodgy investors.

Rather than a product to make money from, we want our dragon to be more like a friend or family member, which could easily happen if we are with them from the start and while they grow up. Have you ever seen the animated movie "How to Train Your Dragon" or any of its sequels? If so, you will know that the plot has a clever twist – the main character, Hiccup, is meant to kill a dragon at some point, but instead bonds and makes friends with one. In time, the dragon becomes "his" dragon, which he names Toothless.

How is that possible?

Hiccup finds the injured Toothless and somehow cleverly "MacGyvers" (slaps together a fix to a problem with whatever is available[i]) a fix to the dragon's broken tail. Over time, Toothless and Hiccup become like family to each other. Incidentally, Toothless has plenty of teeth, but since they are retractable, Hiccup gives him that name. Our dragon could have retractable teeth as well, but our task is already a big enough challenge overall, so we haven't decided whether to bother with fancy tooth options. We do definitely want our dragon to have impressive fangs, however, and poisonous ones would be a bonus.

A funny side note is that the word "Pteranodon" as a scientific name may be translated literally as "toothless wing." We wonder if the creators of *How to Train Your Dragon* were aware of that meaning when they named the main dragon character Toothless.

Our hope is that by creating and raising our dragon, we'll build a kind of familial relationship with it, just like the one between Hiccup and Toothless. The dragon should see us in a positive light and to develop a tight bond with us. But then again, sometimes kids grow up to not be exactly 'fond' of their parents. Plus, our dragon will be real, unlike Toothless, and also, we cannot find an injured dragon to fix as a means of bonding. If only we could, then we could try to clone it.

---

[i] https://en.oxforddictionaries.com/definition/macgyver

The main point here is that – unlike Hiccup or *GoT* characters – we cannot just go out and find dragons or their eggs.

So. Back to reality. This means that we'll have to build a dragon or even better, a breeding pair, or even a bunch of dragons, despite the risks. Recent advances in genomics, gene-editing with CRISPR, bioengineering, and stem cell technologies, combined with some bold ideas and more than a dash of luck, might just do the trick.

Or we might die in the process, be arrested, have our dragons stolen by the CIA (or some other spy agency or the military or just some random person), or we might succeed, only to have the dragon turn on us and kill us – in any number of ways – without warning. Doesn't this sound like a lot of fun?

But, believe us, if we succeed it'll be worth it. To us, the project sounds like an amazing quest.

What is a dragon exactly?

Before we can build a dragon, we need to ask, "what is a dragon exactly?"

Dragons, or dragon-like creatures, have appeared in the mythology of almost every culture around the world, in some cases for millennia. The type of dragon that we see in most movies, and that some of us picture in our heads, is a specific "European" type of dragon. It breathes fire and has wings with which it flies. It also has scales, spends most of its time on land when not flying, and carries mostly evil intentions (which we would try to engineer out, at least towards us, but we're not quite sure how). However, dragons in many other cultures bear a strong resemblance to serpents, while still in some cases sharing certain similarities with Hollywood-style dragons. Many of them aren't evil.

So, where did we start with our dragon research while writing this book? It seemed only right to begin our historical dragon journey in some people's notion of the cradle of civilization itself, Mesopotamia.

The Middle East and Africa

In ancient times, in what is now the Middle East, there was a region called Mesopotamia (mostly in Iraq today), in the south of which was an ancient land called Sumer. The people of Mesopotamia and Sumer are thought to have believed in a variety of creatures with dragon-like features. Some were serpent-like creatures, but sometimes these "dragons" had bird-like features or were lionish in appearance. There was also something of a combination described in ancient writings there specifically called a lion-dragon.[ii] These dragons overall had great powers.

While these creatures were like our modern conceptions of dragons in some ways, they also had many distinct features too. For instance, they didn't breathe fire, but storms instead, which we think is a fantastic idea to at least consider for our dragon (although we'll likely stick with fire). These beasts were generally thought to bring very stormy weather. Ancient Egyptians had their own types of "dragons" as well including one named Apep, who like the Mesopotamian dragons was associated with stormy weather, but also with additional phenomena like earthquakes and eclipses.[iii] Other Middle Eastern cultures had their own versions of dragon-like creatures too.

Asia

Dragon-like creatures were also associated with storms elsewhere as well. In Bhutan and Tibet, the Druk, a thunder serpent, is still a well-known mythological creature that bears some resemblance to and is largely thought of as a dragon. In fact, starting a few hundred years ago Bhutan as a country started sometimes going by the name "Druk Yul" or "Land of the Thunder Dragon." Also, the dragon Druk is on the Bhutanese flag today.[iv]

[ii] http://oracc.museum.upenn.edu/amgg/listofdeities/ikur/
[iii] http://allaboutdragons.com/dragons/Apep
[iv] http://allaboutdragons.com/dragons/Druk

Figure 1.1. Embroidery depicting Japanese dragons with peonies. Note that these dragons don't have wings or breathe fire, but they do have four legs. Image from Shutterstock.

In ancient Hindu text, things are flipped around somewhat as a big serpent-dragon named Vritra sometimes caused droughts rather than rain storms. It was thought to cause droughts by literally hoarding all the water.[v] Vritra is one of the many foes said to have been defeated by the king of the gods, Indra.[vi]

Japan, China, and Korea have had important dragons in their cultures for thousands of years. Japanese dragons were water deities, which were wingless. They lived near and in both rivers and lakes (Figure 1.1). Chinese dragons were also wingless and associated with water; more specifically, with rain. There are parallels with the Indian dragon Vritra here too. If the harvest wasn't good in some parts of China, due to droughts, the people thought dragons might be

---

[v] http://allaboutdragons.com/dragons/Vritra

[vi] https://www.britannica.com/topic/Indra#ref942544

somehow involved and would be able to summon rain if the people paid tribute to them.

Dragons were seemingly more revered in these cultures rather than thought of as evil monsters as they are viewed in many other places around the world including in Europe (see below). In fact, some Chinese emperors claimed to be incarnations of divine dragons, and dragons took on royal and divine meanings.

Many Chinese villages used to build long (as long as three or more people standing one on top of the other) cloth and paper dragons, which they would use in dances, performed during harvest festivals to bring rain. A lot of villages would also hold dragon boat races. The dragon remains a major figure in Chinese culture now. It is the fifth animal on the Chinese zodiac. Some Chinese believe that people born in the year of the dragon are more likely to be powerful, brave, innovative and strong. To this day, various celebrations in China often involve dragons.

Ancient and Eastern Europe

There are also dozens of references to monstrous, serpent and snake-like creatures in Greek and Roman mythology that are reminiscent of dragons. All of these creatures are wingless, watery, and more or less nasty. One of the earliest accounts of a Greek dragon was of a blue dragon called a drákon. This dragon adorned the armor of the famous Greek king Agamemnon (you might have heard of him – he commanded the Greek army in the Trojan wars and features in the famous poem called *the Iliad* as well as in the movie *Troy* starring Brad Pitt).

Drákon. Sound familiar?

Although we aren't historians and could have missed something, this was the first example we could find of a serpentine beast being called something that resembles the word "dragon."

And then there were the "zmey" or "zmeu", which in Slavic and Neo-Roman (Russian, Romanian, Italian, Albanian, etc.) mythology were large, three-headed snakes that breathed sulfuric gas and fire.

These were the first dragon-like creatures that we found to breathe fire, and which more closely fit with our own idea of a dragon.

Central and Western Europe

Dragon myths were ubiquitous in Central and Western Europe. These dragons were winged, serpentine, fire-breathing and, of course, evil. They were perhaps first mentioned in Norse mythology. The first ever Norse dragon, named Nidhogg, was thought to have gnawed at the roots of the "world tree"[vii], which sounds impressive to us. Thor also battled a tremendous dragon, representing a common theme in many parts of the world where heroes or gods battle dragons and as a result become even more heroic.

One of the most famous Western myths about dragons is that of St. George and The Dragon (see the painting depicting this battle in Figure 1.2).[viii] There are numerous versions of this myth. In one, a dragon kept terrorizing a kingdom in Libya and even killed a young shepherd. To keep the dragon at bay, the people sacrificed two goats to it every day. However, the dragon soon grew impatient and required the people to sacrifice their own children. Finally, no one was left except the king's own daughter. She was chained to a rock, in a bridal dress, near the lake where the dragon lived. But before she could be eaten, the knight St. George saved her and killed the dragon. This kind of stereotypical theme of a tough "good guy" saving a "damsel" from a bad monster or person permeates some art such as movies even to this day, even if they have nothing to do with dragons.

These dragon-killing exploits of George occurred before he was so famous or a saint, and played a big role in the growth of his image (imagine if all of this was on Twitter). They were noticed by King Richard III in the 12th century. These heroic deeds appear to have been the main reason Richard chose George as the new patron saint of

---

vii https://mythology.net/norse/norse-creatures/nidhogg/

viii http://www.bbc.co.uk/religion/religions/christianity/saints/george_1.shtml

England (known then as Anglia) displacing the previous patron saint, St. Edward.[ix]

Figure 1.2. St. George killing "the dragon," which in this case is a surprisingly small beast that perhaps – at least as depicted in this painting – couldn't breathe fire but had wings (which are way too small to support actual flight). Image from Shutterstock.

[ix] https://www.historic-uk.com/HistoryUK/HistoryofEngland/Edmund-original-Patron-Saint-of-England/

We kind of doubt that anyone would make us patron saints of anything for building dragons.

Note that in artistic impressions of St. George's foe, the dragon is surprisingly small (Figure 1.2) and there's no sign of it breathing fire. In some accounts of St. George, the dragon's weapon was poisonous rather than fiery. (Talk about bad breath!) It seems that this dragon, which was a wyvern (a dragon with wings and only two legs), was more symbolic of evil than anything else.

One of the most common myths about dragons in Europe overall is that people would supply dragons with food in the form of farm animals like cattle. If this was no longer possible or the dragon became unhappy, then the dragon would start to eat the people of the nearby village instead. We have given some thought to what to feed our dragon (more later in the book) and we hope it doesn't turn to nearby farm animals or villagers.

Dragon history: the big picture

So, what did we learn from this research as a whole?

From this historical perspective, we captured some common themes that we kept in mind when planning our dragon building:

> 1. Practically all mythological dragons are serpentine and associated with bodies of water or rain.
> 2. There are lots of three-headed European dragons, but not two or four-headed dragons for some unknown reason. In other cases, dragons had many, perhaps innumerable heads.
> 3. Most Middle Eastern, Proto-Indo and Southern Asian Dragons are associated with thunder and storms (or drought) rather than with fire, although lightning is often in the mix.
> 4. Only European dragons are associated with fire and it seems that predominantly Western dragons have wings.

5. In Europe, dragons are viewed as evil and often must be fed or given human sacrifices, while in Asia dragons are more revered as powerful and wise.

We're American and mostly of European descent ourselves. Furthermore, most of the media we view is very Western in nature. As a result, our concept of dragons is somewhat limited to our Western culture and, in turn, we are focusing on "Western" dragons in this book, but it could be a riskier strategy because Asian dragons would be relatively easier to build. No fire or wings to worry about. And although a huge portion of the world views dragons as big water snake gods, we want to make flying and fire-breathing lizards.

So it seems that we're mostly set on making the classical European dragon. But it's been good to learn about what other people around the world imagine when they think of a dragon. If we get good at dragon building, we could even branch out and try to make the dragons of other cultures as well. We could even take a leap and attempt to make other mythological creatures, such as unicorns. In Chapter 7, we talk more about making such creatures.

In a way, what this history is telling us is that we don't have to conform to Western or European ideas of what a dragon should be like and that we could be creative in some ways. Also, if we get stuck on some tougher challenges, like fire-breathing, we could opt for a more Asiatic kind of dragon, which would still be cool.

Why build a dragon?

Despite all of our research, some of you might still be asking: why try to build any sort of dragon in the first place?

Well, we think there are lots of reasons.

What could be more of a thrill than having your very own dragon with you as you go about your life?

If it was big enough, the dragon might not just accompany you but also fly you around the world. You could ride on its back, à la Daenerys Targaryen of *GoT*, but instead visit real places on amazing trips with your dragon (for example, from London to Las Vegas or Bangkok to Johannesburg) rather than just across the imaginary *GoT* land of Westeros on TV.

Now the skeptics amongst you might say: where would your dragon land on these trips and where would it hang out safely? We've decided to prioritize making the dragon first and then worry about that stuff later – if we're still alive.

A real dragon could also bring other benefits to us, as its creators or "owners" (or to you, if you follow in our footsteps or buy a dragon from us, not that we are in this for the potential payout so maybe the money could go to charity). But it's not easy to imagine ever owning something like a dragon, a creature that might not like being owned and that could easily do something serious about that feeling. So why else would it be great to have a dragon?

Not that we would do this, but – in theory – our custom-made dragon could cremate our enemies (or yours, if you make one too) or at least scare the hell out of them by threatening to do so. What better way to become tremendously powerful than to have your own dragon incinerate a Hummer or ferocious wild boar, charging at you with its tusks out, turning them into wispy clouds of carbonized dust with just one breath?

One could imagine this demonstration going down while crowds of people look on, amazed and awestruck. Or at least that's how some folks might envision it, not that we would. Even without illusions or delusions of grandeur or gratuitous violence, a homemade dragon would certainly be a much more interesting companion than a poodle, domestic cat, or cockatoo.

If we can build one dragon, we could, over time, probably put together a flock or flight (or whatever a group of dragons is called). If we can successfully make a group of dragons that don't kill us, what should we call them? A flock of dragons and their riders were called a "weyr" in Anne McCaffrey's fictional world of Pern, but frankly, that's not terribly exciting as a name. A more suitable name would be a "murder" of dragons, borrowed from the collective noun used to describe a group of crows.

The phrase "murder of dragons" has a nice ring to it.

If we master the dragon-building process, we could make a murder of dragons. But instead of building all of them separately in the lab, we could be smart and make a pair of fertile dragons and then let them make baby dragons. Lots of them. Or we could make dragon clones, which we discuss further in Chapter 6.

Either way, we could call the baby dragons by cute names like dragonlings, dragon hatchlings, or dragonlets. We bet they'd be cute, at least until they start to grow up and begin chasing the neighbor's rooster around the yard (and roasting it for dinner) or worse, start chasing the neighbors or even us. You get the picture.

Once you have your dragon or dragons, Tony Stark, er, Elon Musk, might give you a call to hang out. Or Taylor Swift might ask for a lift to Paris, or even for a favor of you and your dragon by way of revenge on some ex.

Bill Gates might suddenly want to have dinner with you. This uber-tech guru could be eager to discuss your dragon plans. Perhaps you might chat about how he could help you to set up a dragon institute with all the space and budget you need to make more dragons. Or maybe, Daniel Radcliffe calls. He's wanting to live out a real-life dragon fantasy trip as a throwback to his Harry Potter filming days. Or, maybe that's just us dreaming…

A real dragon would have something magical about it in people's minds, even if it was a living, fire-breathing animal rather than a charmed, figment of someone's imagination. If someone like us (or you) pull this dragon-making adventure off, they'd maybe be voted the most awesome person on the planet with the hottest pet (in more ways than one). However, a dragon would be more than a pet. We would need to get along with our dragon. And it can't go on a killing spree or get killed itself because it's too dangerous. We don't want a global desire for dragons to turn into a global desire for their extermination. The reputations of some mythological dragons for doing evil would probably be quite unhelpful in this regard.

For us to make this happen – and to not end up rapidly dead ourselves or have our dragon be nuked – our dragon must have certain traits and not go wild. It can't be too wicked in temperament. It can't go off and make its own barbequed brunch of the family down the street. And it definitely can't turn its fire, claws, or teeth on you…or say, all of New York City or Shanghai. We don't want a Godzilla or a King Kong situation here. Our dragon shouldn't invoke a military response, or this adventure could turn into a lot of trouble. On the other hand, we maybe don't want a dragon that is a peace-loving vegan, either.

What kind of dragon should we make?

Maybe we are getting ahead of ourselves.

We want a dragon and there are a million reasons why, but let's back up a bit.

What will this dragon be like?

What specific features do we want our dragon to have?

And how do we build the thing in the first place?

What creatures could we use as our starting point if we don't want to start from scratch by bioengineering a dragon from just cells or by using a 3-D printer? Yeah, we realize that it'd have to be a huge 3-D printer. Or it would have to print out smaller, snap-together, semi-living dragon pieces, which is not very compatible with a biological entity like a dragon (it's not a robot after all).

First, we want our dragon to look and behave like a "dragon" in the sense that people from diverse cultures around the world would each look at it and think "dragon."

When people see it, we don't want them scratching their heads and wondering, "Heck, what's that critter?" They also shouldn't mistake our dragon for something more boring. Here are some questions we don't want to hear: "Is that some weird flying alligator?" or "What's that freaky little lizard up in the sky? A kite?" Ideally, they'd instead instinctually yell – if they can speak at all – "Dragon!" Then they'd either scream and flee for their lives or turn to stone (figuratively) as they stare in amazement and wonder.

Our expectations aren't too high, are they?

These kinds of intense dragon-worthy reactions should happen to onlookers, or we'll have failed. This requirement means that being reptilian in appearance is necessary but, in itself, isn't enough. The total dragon package should be obvious to anyone.

What creatures to start with?

Our ability to create a real dragon with this kind of unmistakable appearance will in part come from making some smart choices about where to begin the dragon-making process. We think that a wise strategy would be to use an existing creature, or a combination of creatures, which in some way bear an unmistakable resemblance to a dragon or at least have some clear dragon-like traits.

There are some real-life creatures in various parts of the world that already bear some resemblance to dragons, and which even have the word "dragon" (in the native language) included as part of their names sometimes. These creatures give us a whole range of options as a dragon starting point. But the two that come to mind first are, funnily enough, starkly different to each other: the first creature is the small Draco lizard (e.g. *Draco volans,* which you have to admit is a cool name) and the second is the huge, sometimes lethal, Komodo dragon (*Varanus komodoensis*).

You can see Komodo dragons pictured in real life in Figure 1.3. Although the human photographer is some distance away in the background, you can hopefully get a sense of just how humungous Komodos are.

Why potentially start with these creatures?

Both Dracos and Komodos are reptiles, so that's a plus, and both have names that suggest that people have, for many years, seen them as being dragon-like. Note that "Draco" means dragon in some languages. In the plus column specifically for using Dracos as our starting material in our dragon-building quest, is the fact that they can already glide great distances in a way that is close to flying. Dracos also kind of look like miniaturized dragons.

Figure 1.3. Komodo dragons on the island of Rinca, Indonesia. In this photo, you can roughly compare their size to that of a human photographer in the distance behind them. Image from Shutterstock.

On the other hand, a potential downside is that Dracos are rather small creatures and so won't likely give the impression of being a big, fearsome dragon to onlookers. If, despite our best efforts, our dragon ends up being that tiny, we'll be reminded of the funny character Mushu, the tiny dragon from Disney's *Mulan*, whom the title character mistakes for "just" a little lizard.

Given their naturally small size, if we start with Dracos, we'd probably want to give them a major growth boost during development (more on this later in the book) to make a more dragon-sized final product. Even small, real-life Dracos have some impressive features, such as flaps called patagia that catch air allowing the lizards to soar, as well as interesting coloring. But they sure are tiny creatures.

Komodo dragons are a very different story.

They already possess cool dragon-like features. For example, they look like dragons and are impressive in size. They are also adept killers, even of humans, and even just a minor Komodo bite can be dangerous [2], which is a nice added bonus for any dragon.

However, using Komodos as a starting point to make a dragon comes with some challenges. The main obstacle is being able to make them fly given their impressive yet bulky size. Without some kind of genetic change to reduce their mass, getting one of these giants airborne (and back down again in one piece) might be the equivalent to somehow getting a full-grown elephant to successfully pole vault 20 feet (ca. 6 meters) and then land without harming itself. Maybe even with a "Ta-da!"

As we discuss in Chapter 8 on ethics, another challenge with Komodos is that there are so few of them left in the wild, so we wouldn't want to endanger them.

So what have we learned from considering both Dracos and Komodos as potential starting creatures for making our dragon? Well, it seems clear that each creature we consider for this will come with a range of both potential advantages and disadvantages. It's going to be a mixed bag in every case.

Chimeras?

One potentially promising "best of both worlds" route to creating a dragon is to make a chimera – a creature that is made up of different species. We could make a chimera by combining the embryos of different animals or by using genetics to combine useful genes (and the "best," dragon-related traits generated by these genes) from different animals to create one new animal. Imagine a cross between a Komodo and a Draco – you might end up with a deadly creature of intermediate size that could soar from tree to tree.

If this seems improbable, look at all the various amazing outcomes of dog breeding. The result of breeding together different kinds of dogs can sometimes be hilarious – such as crossing a golden retriever to a corgi with tiny legs. The result could be seen practically speaking as a chimera of sorts, although it might be more correct to call it a hybrid.

Our envisioned chimera could have desirable dragon-like attributes from different species – not just from Dracos and Komodo dragons but from other animals as well. Two animals that come to mind are, unexpectedly, insects: specifically, dragonflies (for example, the very large *Petalura* species) and the bombardier beetle (*Brachinus* species), which, as mentioned earlier, can shoot a hot, fiery mixture out of its butt.

Figure 1.4. Blue Rhino Studio. Artist's conception of a Quetzalcoatlus. Note how the Quetzalcoatlus is imagined being able to stand (and likely walk) using its wings as front legs. Photo used with permission.

Admittedly, using vertebrates (that is, animals with a backbone) as our starting creature might be more practical than using an insect, but dragonflies might get our dragon flying, and the bombardier beetle's physiology might help us to figure out how to get our dragon breathing fire (more on that later).

Still, we're not sure that insect dragons could be made large enough to look like our imagined dragon, so we'll focus mostly on using vertebrate animals in our dragon-building plan.

Our dragon chimera could even include design elements from creatures that are now extinct, such as pterosaurs or Pteranodons (better known by the public as pterodactyls), including perhaps the largest flying creature to have ever lived, called Quetzalcoatlus (see Figure 1.4 for an artistic vision of what this creature could have looked like). Incidentally, Quetzalcoatlus was named after Quetzalcoatl, a god from Mesoamerica that was like a feathered serpent (yeah, dragon-like!).

We'll talk more about pterosaurs in a bit.

And what about using genetic technology to make a chimera? The basic idea here is that rather than mixing cells or embryos of different species into a new combination, you insert just key genes from one species into cells of another species. This route gives us several options. We could induce the required genetic changes in stem cells or in the reproductive cells (the sperm or egg) of our starter animal, or we could make genetic changes directly in their fertilized eggs. It would be easy to think that we could avoid some predicted problems (dragon too small or too big, etc.) by making genetic chimeras instead of mixing embryonic cells. But making genetic chimeras will very likely cause new problems for us to consider.

Overall, we realize that some of our more outlandish ideas for chimeric combinations might not turn out too well or be particularly practical. To explore these approaches, and their pros and cons, further we have dedicated a whole chapter (Chapter 6) later in this book to explain how reproductive and gene-editing techniques could be used to make dragons including *via* chimera technology.

Flying dragon

What else do we need in our dragon?

Although some mythological wingless dragons, such as those shown in ancient Chinese art (see Figure 1.1), spent much of their time swimming (*allegedly*; we are assuming in this book that dragons aren't real *yet* and never were) – we want to produce a dragon that can spend a good amount of time in the air.

And so, second on our checklist and the focus of our next chapter (yes, you have probably guessed it) is flying. Our dragon must have wings and fly. In this book, we explore how to engineer dragon flight (and how great or awful things might turn out on this front). Along the way, we introduce you to how birds and other creatures are already great at flying. Think about it. How exactly do flying creatures accomplish flight and how does their physiology make it happen? It's not so obvious and it makes for a pretty interesting story in Chapter 2.

With that goal in mind, as we engineer our dragon, we'll have to both give it wings and keep its overall weight relatively low. One helpful way to achieve that is to give our dragon light-weight but strong bones, perhaps ones that are hollow, like the bones of birds. We'll also have to aim for specific bone lengths in the wings and digits, to make flapping and flying aerodynamically possible. These features, combined with powerful pectoral muscles, should give our dragon the ability to fly.

What about feathers?

We'll keep that option open as it could be helpful, and it could give our dragon an interesting, colorful appearance.

This train of thought also raises the possibility of using specific birds as our starting animal on the path to dragon-hood. But again, we want our dragon to be impressive in size, so we're not aiming for a robin or a hummingbird-sized flying dragon. Size matters for dragons if we want it to fly (which we do), in the sense of not being too big nor too small. There'll be a sweet spot to hit when it comes to size but that target size should be impressive (if all goes well). Still, starting in the avian world is a possibility, and a particularly large bird could be a starting point worth considering.

In our view, Pteranodons like Quetzalcoatlus, which are extinct relatives of birds, were the real-life animals most similar to dragons. They were huge, lizard-like animals, which seemingly could fly based on research by paleontologists. We can just imagine them soaring above terrestrial dinosaurs and other earth-bound creatures of the planet at the time, and the image that comes to mind is in some ways dragon-like.

There's a lot we can learn about building a dragon from specific types of Pteranodons since they managed to be giants and still fly (even if they didn't breathe fire). An interesting side note is that historical finds of fossilized dinosaur bones, including those of Pteranodons, might have inspired some dragon legends.

In the next chapter on flight, we also discuss how it is that birds can fly, how bats evolved flight even without feathers, and how wingless animals, such as Draco lizards and flying squirrels, manage to get airborne (if temporarily) in ways that we humans can't without considerable help (such as wearing Iron Man's rocket suit). The most common flying animals on the planet right now are insects. Again, we don't think we'll make an insect dragon, but surely there are things we can learn from flying insects, including from dragonflies, that will help us to build our real flying dragon too.

If we had hundreds – or more likely thousands – of years at our disposal, we could try to slowly evolve a dragon from existing creatures or perhaps from an initial chimeric creature. This might have the added benefit of allowing our created dragon to adapt to real-world conditions. After all, another challenge facing us – even if we manage to successfully engineer our dragon – is this: what if it is incompatible with our world? And dies? However, we don't have all that time, and we want to still be around for a long time with our dragon.

(By the way, if you want to know more about flight, and about how evolution has generated many amazing kinds of winged creatures, we recommend reading about wing and flight evolution in Carl Zimmer's book, *The Tangled Bank* [3].)

Fire is fun for a dragon

Finally, our dragon needs to breathe fire or where's the fun? In Chapter 3, we explain how we plan to give our dragon fire.

There are many challenges to making our new beast a fire-breather. For instance, where's the fuel? We shouldn't have to shovel coal or kindling into its mouth or hook it up to a propane tank around its neck. We need our dragon to readily supply its own fuel for its fire. Our ideas for a dragon with a self-sustaining fuel source range from it being able to produce its own range of gases, such as propane (think explosive burps), to more extreme ideas, such it being able to produce nearly pure hydrogen gas. Producing concentrated hydrogen gas, which is very explosive, would be extremely dangerous.

To turn fuel into fire, we also need a spark, and we have some ideas about how this could be achieved as well, such as the electric eel idea. Our dragon could also have other ways to light its fire – it could have flint fillings in its teeth, or we could even give it some cybernetic upgrades to spark ignition. Then we could also cheat but giving our dragon a lighter or matches, but that wouldn't be as impressive. Still, every now and then in this book we mention potential cheats like this to get us over roadblocks we might encounter.

Another option for fire is to generate various reactive chemicals inside our dragon's body but in separate places. These chemicals could then be mixed on demand to generate explosive, fiery reactions – à la the bombardier beetle – but taken up several notches in intensity and coming out of its front, rather than its back, end. However, we don't want our dragon to burn itself from the inside out or to explode so it'll need some internal safety features as well. One way to keep our fire-breathing dragon intact could be to use a protective structure like that which bombardier beetles already have. This internal structure keeps them safe from the nasty, near-boiling point chemical reactions that happen inside their bodies and that feed their explosive "farts."

Our dragon's diet is going to be key as well. The various foods our dragon eats will impact its fuel production. Also, our dragon will need to remain svelte in order to fly but will likely require tons of calories, as both flying and fire-breathing require a huge amount of energy. In addition, this means that our dragon's metabolism is going to be quite important. Maybe it'll need a personal chef and dietitian too? If that sounds expensive, consider that if we are going to make a dragon, we can't really penny pinch, right?

To achieve each of these goals, we can use a combination of technologies, including stem cells, assisted reproduction, CRISPR-based gene-editing techniques, and bioengineering. And as we were writing this book primarily in 2018 and that year was the 200th anniversary of Mary Shelley's *Frankenstein*, we thought of an entirely different approach to building a dragon. We could jimmy together various parts of other animals to form a Frankenstein dragon (of a sort) but it seems likely that it could end up being more monster than controllable, cool creation. We admit that kind of disastrous outcome could still occur, even if we don't go for the more-radical Frankenstein construction approach.

Brain and temperament

From all of this, we are already fairly confident that we could endow our dragon with the physiology it would need for flying, but then how do we teach it to fly? And even if we do, honestly, we aren't so sure about how it will learn to come back and land in one piece. Imagine you build a dragon, show it how to fly, but then forget to teach it how to land or it just never quite gets the hang of it. And then, when it "leaves the nest" for its first real, high-flying adventure, its return ends in a gory "splat" as if someone had catapulted a cow skyward and gravity brings it back at high speed.

For these and other reasons, as our third goal, we want our dragon to be on the moderately smart side. It must be fairly brainy but not just in an emotionless, computer-like way. It needs to have the consciousness to respect us, so it naturally follows our instructions, and doesn't roast and eat us. It must have the intelligence to do many things well beyond landing after each flight.

Still, we aren't necessarily aiming for a dragon that's overly loaded with brains as that could cause other problems, such as difficulty with flight (big brain to carry). We also don't want to end up with a dragon that views itself as too smart to need us around. Never a good thing with dragons. But a dragon that is dumb-as-a-doornail could be disastrous. For instance, how would we teach it important stuff – such as how to fly, how to land, and not pillage the nearby village – if they have the IQ of a turnip?

This brings us to our fourth item, which is another brain-related trait. Our dragon must have a certain temperament. It can't be too quick to anger, or we'll soon be the ones turned to carbon dust by its fiery blast. It must be trainable and come to see us as its family. We'll be its mom or dad. At least, we hope so. Engineering a certain temperament may be one of the most difficult challenges of all, as personality is a complex trait that is likely to be controlled by hundreds of genes, making genetic approaches unlikely to be super helpful, as well as by environmental factors, such as how its raised.

And let's face it, when we consider our possible starting creatures, they aren't exactly loaded with gregarious, or even with moderately endearing, personalities, right? Lizards? Birds? Well, maybe some birds have attractive traits, such as affection and parental instincts, and some can talk up a storm, just like people.

Also, life would sure be a lot easier if our dragon could talk and either speak our language (English) or some unique language that we could understand and use to talk to our dragon in secret. Of course, talking requires certain brain, airway, and mouth structures, but again, since birds can speak very well, we don't think this will be a big problem were we to start with birds. On the other hand, were we to start with lizards, it could get far more challenging.

Getting back to personalities, by contrast to lizards or birds, think about dogs as a species. They are not only highly variable (in science we call this "polymorphic") in how they look, but they also differ dramatically in their personality and temperament. Compare, for example, a Pitbull or Doberman to a Yellow Labrador. (We'd want our dragon to lean more toward the Labrador side, personality-wise, but there are a lot of canine personalities).

In contrast, it's hard to think of any particular reptile out there that has a particularly cheery or attractive personality that we could use as a starting point. However, there are some seemingly happy, mellow birds out there, so that is a potential alternative starting point. But then again, if we think of Alfred Hitchcock's *The Birds* (a horror movie in which birds mysteriously begin to attack people and the people have surprisingly little ability to stop them) and replace the marauding birds with dragons (even little ones)….well, it is far more terrifying to imagine.

We devote all of Chapter 4 to considerations related to our dragon's brain. One of the big challenges on the brain front is that aiming for one desirable trait here might always come along together with a "bad" trait. The brain does not just determine intelligence, but also personality and feelings. As such, boosting intelligence might have negative effects on personality, and *vice versa*. We don't want a dragon who is a sociopath or psychopath on our hands; we would also feel quite guilty having introduced them into the world.

Our dragon-building team

Thinking about fire, will we need our own fire department or at least one firefighting expert to help with avoiding our dragon burning down all kinds of buildings? Probably. The dragon could be especially prone to fire-breathing accidents when it is young. If we start our dragon-building process with birds, it might be wise to have an ornithologist (bird expert) on the team. If we start with a lizard, we'll need a herpetologist.

Come to think of it, while we have a lot of bright ideas and Paul has decades of experience as a researcher, we are likely going to need a large team to successfully build our dragon. We'll also need many other kinds of biologists and maybe chemists, computer experts perhaps to help with artificial intelligence AI-based design, reproductive specialists, veterinarians, and more.

And, yes, we realize costs are quickly going to add up for such a team so later in the book we discuss how to raise all this money.

Expect disasters

Speaking of bad outcomes – to be fair and because it is kind of interesting – throughout the book we also discuss what could go wrong and when, and how we might die trying to make our dragon. We imagine this might be kind of akin to Alfred Nobel and his team going through the process of inventing dynamite, knowing that at any point along the way, things could go very wrong. By the way, Nobel's younger brother Emil died due to an experiment that went wrong. Serious science, like dragon making, has real risks too.

But let's say that we achieve all of these goals– to build a dragon with certain traits – we still wonder how the world will respond to our new dragon (or perhaps "murder of dragons" if we succeed in making a group)?

If security agencies worldwide, such as the NSA, MI6 or FSB (the newer incarnation of the old Russian spy agency the KGB), or any other powerful spy agency, comes calling, we don't want them to view the dragon as a loose cannon that must be eliminated. They might also come knocking on our door wanting to take our dragon off our hands to use as their weapon. They might also want us to produce a bunch more dragons for them to use for various purposes to "benefit humanity" (yeah, right). Imagine dragon soldiers.

We haven't yet figured out how we could deal with spy agencies, or any other organization, trying to nab our dragon or our dragon-making technology. Should we perhaps patent or copyright our dragon methods and our actual dragons?

And what about the public? The public might be both fascinated and freaked out by our dragons. Some people on seeing our dragon might recall the religious story of St. George killing the dragon, and think "evil!" When you throw in the fact that dragons are often used as a symbol of evil and can evoke terror across the world, our dragon might not have many fans in certain circles.

Then there's also the ethical side of all of this. Would it be ethical to make a dragon or even to try to make one? What factors come into play as you mess around with making chimeric or genetically modified creatures? We cover such questions in Chapter 8.

Finally, we would need a large amount of funding to make our dragon a reality. Most likely this means our dragon research being funded by one of three possible sources: investors, a single rich funder, or some kind of online fundraising effort. In any of these three scenarios, problems could arise. Maybe we could launch a campaign on GoFundMe? It might raise a lot of money, but there goes our secrecy out the door.

It is also possible that a government could supply the money, such as the U.S. government or its defense funding agency, DARPA. However, in this scenario, we would again worry about the dragon being used as a weapon. Since we could need just a million dollars to get started (even if that would rapidly grow to ten million or more later), we hope that to get going in such a way that we remain in control of the project (Jurassic Park had investors and look how that went.)

Why write this book?

We know why someone would try to build a dragon, but why write a book about trying to build a dragon?

First of all, we thought it would be fun to see if we could figure out how this project could actually be accomplished. Could we come up with a dragon-building plan that wasn't wildly improbable? We think we did, and there are many alternative approaches that we could opt for should roadblocks pop up along the way.

Our second motivation was that we figured that we'd learn about a lot of cool science that we didn't already know as we researched our plan, which we could pass onto others, in particular kids and young adults but also to those who are just young at heart.

Another motivation to write this book was satirical. For example, we wanted to poke a little (okay, a lot of) fun at science and the media for overselling new research. In this way, we thought we could convey just how often exciting science gets hyped. For instance, although CRISPR gene-editing could be uniquely helpful for building a dragon, it has become so hyped as a gene-editing technology that some people seem to think it can be used to achieve any kind of comic book-like goal. Such hyping of CRISPR makes it an easy target for us to poke fun at as we discuss making our dragon. The same goes for other technologies that we might want to use, such as stem cells and cloning.

What this means is that – as you read – you might bump into a hyperbolic statement or two. Please interpret these statements as we intend – as being ironic or satirical in nature. We won't warn you again. Well, we might, but hopefully, you get the picture just from this heads-up. Also, we suggest reading the Preface.

It's also worth noting that many other folks have speculated how they could make a dragon, including writer Kyle Hill at *Scientific American*, from whom we got some inspiration, including about a few possible ways to engineer a dragon to breathe fire.[x] A hat tip to Kyle is in order for his ingenious ideas.

Is it even possible to think about how to make a new creature that is remotely like a dragon using genetic technology? And, if so, what about creating more strange animals, such as unicorns? Or tiny elephants? Others have written about these kinds of questions too, including Stanford Law Professor Hank Greely and ethicist Professor Alta Charo [4], as well as various journalists.[xi]

Greely and Charo seem much more skeptical than we are, but even they acknowledge that someone might attempt such a project and get a long way toward their goal. In an article published in 2015, they argued that something like a dragon could be produced:

> "Basic physics will almost certainly combine with biological constraints to prevent the creation of flying dragons or fire-breathing dragons – but a very large reptile that looks at least somewhat like the European or Asian dragon (perhaps even with flappable if not

---

[x] https://blogs.scientificamerican.com/but-not-simpler/smaug-breathes-fire-like-a-bloated-bombardier-beetle-with-flinted-teeth/

[xi] http://www.bbc.com/news/uk-wales-35111760

flyable wings) could be someone's target of opportunity."

Other than some spy agency or the US defense department taking our dragon from us or buying the technology from us part way through our project, wouldn't some other US government agency, such as the USDA, EPA, or FDA, step in to stop us? Maybe, but government agencies are notoriously more reactive than proactive, so we might just go for it and hope for the best. We discuss regulatory agencies more in Chapter 8 on bioethics.

With this introduction, we hope that you'll join us as we go further into our journey into how we would create real, live, fire-breathing dragons.

References

1.   Pickrell, J., Huge haul of rare pterosaur eggs excites palaeontologists. *Nature* 2017. **552**(7683): pp. 14–15.
2.   Borek, H.A. and N.P. Charlton, How not to train your dragon: a case of a Komodo dragon bite. *Wilderness Environ Med* 2015. **26**(2): pp. 196–199.
3.   Zimmer, C., *The Tangled Bank : An Introduction to Evolution.* Second edition. ed. 2014, Greenwood Village, Colorado: Roberts and Company.
4.   Charo, R.A. and H.T. Greely, CRISPR Critters and CRISPR Cracks. *Am J Bioeth* 2015. **15**(12): pp. 11–17.

Chapter 2

Let there be flight

What's a dragon that can't fly?

To us, it would feel like a big letdown.

As discussed in the last chapter, throughout history, it is true that there were plenty of mythological dragons that didn't take to the skies – they were either earthbound walkers or lived in the sea. Yet, they were still amazing creatures and were probably not often mistaken for being anything but fierce dragons. However, in this book, the goal is to make a flying dragon, if at all possible.

If we fail to give it flight, our dragon would still be a ferocious, non-flying beast that could maybe run rather than walk. In this way, such an earth-bound dragon could be like a roadrunner (a type of tall predatory bird that runs very fast, but doesn't fly; incidentally "The Roadrunner" is also a major character in some old Looney Tunes cartoons), but a massive one that scoops up its prey and breathes fire. In fact, it's hypothetically possible that only some of what are assumed to be the most awesome flying creatures ever, Pteranodons, such as the Quetzalcoatlus, were able to truly fly (if you need a reminder of what a Quetzalcoatlus might have looked like, see Figure 1.4 in the last chapter, and also the fossilized skeleton in Figure 2.1). We'll talk in more depth about Quetzalcoatlus later in this chapter.

Would it be such a bad thing if our dragon wasn't able to fly? Well, yes. Without flight, our dragon would be limited in its powers and ability to defend itself. And to many of us, who are used to seeing dragons as they are depicted in modern media, it just wouldn't feel like a dragon.

Maybe we are being greedy because we want the whole dragon package, including flight, but can you blame us? If you are going to go to the trouble of building a dragon, why not go for the most amazing dragon possible, right? We suppose we could make different kinds of dragons: ones in the sea, those walking on land, and then the fliers. However, making even one dragon is a challenge, so we've put our energy into planning how to build just the flying version.

Even giving our dragon the ability to fly at a basic level will be a tough challenge. However, our dragon needs to do far more than just fly like some novice pilot puttering haphazardly around. We want it to be at home in the skies and to fly with both speed and grace.

Weight *versus* flight

In order to fly, at a minimum, our dragon needs the power to easily carry its body mass. This requirement alone means that its body mass must be kept as low as possible. The more massive a creature is, the less likely it will be able to fly. It seems that the evolution of flying creatures, such as Pteranodons, and also of birds, which are kind of modern flying dinosaurs themselves, stuck to this principle as well. It's possible that Quetzalcoatlus was the largest flying creature ever to have roamed the Earth, but even it may not have weighed as much as you might think, based on its size (meaning its dimensions).

You may well respond to our idea – to create a flying dragon – with 'it'll happen when pigs fly'. But all we need to do is make a lizard fly, and in fact, some lizards are already close to flying, so maybe that's not as impossible as it first sounds. By the way, making most average-sized lizards fly would be less difficult than making pigs fly. We could also go the opposite route by taking a flying creature that is already related to dinosaurs (yes, we are talking about birds again!) and making it seem more reptilian and hence more dragon-like.

No problem, right?

Figure 2.1. A restored Quetzalcoatlus fossilized skeleton, Houston Museum of Natural Science. Quetzalcoatlus is a type of pterosaur with wings that is thought to have been able to fly. Photo by Yinan Chen (www.goodfreephotos.com (gallery, image)). Image in the public domain *via* Wikimedia Commons.

Okay, we admit that whether we start with lizards or with flying creatures, it's going to be hard to make a flying dragon. We are going to face problems along the way.

Why?

The features we need to design into our dragon to make it a real dragon will tend to increase its mass. For instance, specific gastrointestinal structures that allow for fire-breathing (see more on this in the next chapter) could make it heavier and make the possibility of its flight more remote.

Also, the stronger and more resilient our dragon is, the more it will weigh, due to increased muscle mass and skin thickness. Once again, that source of extra weight will make flight harder too. If you look at fictional dragons, most don't have wings that are anywhere near large enough to realistically power their flight, given how enormous their body mass tends to be depicted. Of course, *this book* is all about sticking to reality as we build our dragon. Or at least we shouldn't be breaking the laws of physics too badly.

What other considerations come into play when it comes to flight? We wildly speculate for a moment in the next chapter that storing certain lighter-than-air gases (such as hydrogen) in our dragon's gut to fuel its fire could make it less dense overall and hence more able to fly. But we aren't looking for our dragon to resemble a hyper-inflated balloon or to be highly explosive.

It's also not just a matter of the dragon having the right power-to-body mass ratio to fly because it needs to be aerodynamic as well. How many times have you made paper airplanes that should fly but instead crash straight into the ground? We've made many such planes, and we don't want our dragon to crash like they did because then we would have to start over, which wouldn't be as simple as folding a new piece of paper into an airplane. And to be honest, we expect to get emotionally attached to our dragon.

In addition to having a certain body mass and powerful wings, our dragon will also need the required brain power to learn to fly, to get around up there, and to have the skill to land in one piece (see more on engineering our dragon's brain in Chapter 4). While it's not entirely clear from research just how much of a flying creature's brain mass is devoted to the process of flight, it seems likely that a high proportion of their gray matter is needed to fly skillfully and land.

And as we've said already, we really don't want our precious dragon to go splat!

Figure 2.2. Draco lizard (e.g. *Draco volans*). Imagine this guy, but with patagia and arms fashioned into dedicated wings, being much bigger overall, and breathing fire. Creative Commons image.

Imagine building a dragon, perhaps spending millions of dollars and years of your time on it, all the while becoming quite emotionally attached to it, only to have it fly off into the sky and then fall, fatally, to earth. It wouldn't even have to plummet to the ground at some amazing speed to end up dead. It could simply try to land but fail by accidentally flying into a tree or building, or landing so hard that it dies. And if we are riding on its back at such a moment it would kind of take some of the fun out of the whole project.

Engineering flight: where to start?

When it comes to the weight *versus* power issue, we can look at some of our potential "starter creatures" to see examples of this dilemma in the real world. For instance, compare the positively tiny Draco lizard to the massive Komodo dragon (also known as the Komodo monitor) or to a giant crocodile. While the crocodile is the largest living reptile, the Komodo is the largest living lizard. It's hard to imagine either of these giants becoming airborne.

(By the way, who knew that reptiles and lizards aren't necessarily interchangeable terms? Now we do.)

The average Draco lizard is thousands of times smaller in body mass than a Komodo dragon, so it's nowhere near dragon-sized. But Dracos are the closest thing to a flying lizard in the world today, and they have a distinctly dragon-like appearance (Figure 2.2) despite their small stature. While Dracos are only about 8-inches (just over 3cm) long and weigh just under an ounce (or roughly 20 grams), a Komodo dragon can be 12-times longer, with a reported record weight of around 366 pounds (ca. 166 kg), which is about 8,000-times bigger than a typical Draco.[i] On this level, which seems more dragon-like to you?

Even so, Dracos can almost fly, while Komodos are far from achieving that. Dracos use skin flaps, called patagia, to glide in the air from tree to tree, covering distances as much as 100 feet (ca. 30 meters) between trees. You can imagine that by using certain genetic or stem cell techniques, we could give the Draco larger muscles, longer bones, and turn its patagia into larger, functional wings. Since these additions could add weight to its small body mass, making it harder to fly and soar, we shouldn't go overboard on wing-size. Even so, our modified Draco could perhaps fly, rather than soar, without other major physiological alterations and without violating the laws of physics.

Throwing in some genetic tweaks to give lighter and longer bones, such as those of some birds, might also help. However, a genetically modified Draco would still likely be smaller than a Komodo, which could make our Draco-based dragon less impressive.

Still, perhaps as mentioned earlier we could instead make a large army of small, fire-breathing Draco-dragons that could coordinately breathe fire together and attack in numbers, like the avian maniacs in Hitchcock's *The Birds*. But we think it more likely that we'd need to use additional technological tricks — such as adding in extra growth factors — to make our "Draco-dragons" much larger and somehow still able to fly.

---

[i] https://nationalzoo.si.edu/animals/komodo-dragon

On the flip side, the Komodo dragon is huge. It's an impressive, deadly beast that can sometimes kill people (see Komodos chasing people in Indonesia in Figure 2.3, and then imagine these Komodos with wings that can fly, and the outcome could get much gorier, but also more dragon-like, which is good news from our perspective).

Komodo dragons regularly take down large prey even without having any laboratory modifications. And they naturally have the size and fearsomeness that we associate with a dragon so their "dragon" name already makes good sense. Practically speaking it'd also be a good animal to start with in our pursuit of making a large, viscerally impressive dragon.

Figure 2.3. Tourists running from Komodo dragons that are chasing each other in Indonesia. Imagine if these were winged, fire-breathing Komodo dragons that could fly. Photo credit, open source image from Jorge Láscar.

It's a bit hard to imagine Komodos in flight like a pterosaur (see fossilized pterosaur skeletons in flight in Figure 2.4). How the heck would you even get a Komodo dragon off the ground to fly, short of a catapult (maybe with a parachute to avoid the potential "splat problem")?

To get our Komodo-based dragon airborne – reliably and safely, and back on the ground in one healthy piece as well, given its mass – it would need to have wings the size of a medium-sized airplane. For its wings to be smaller, we would have to make other changes, such as giving it lighter bones and a more streamlined musculature. The list goes on.

So, we have our work cut out for us, whether we use a small or large creature as a starting point to try to build our new dragon. Still, when we think of our ideal dragon, we want it to have the ferocity and bulk of a Komodo, or close to it, and to have an even greater ability to kill (sorry, but a dragon needs to kill even if it never ends up using that ability on humans). We also want it to have the flying skills of an agile animal, such as a bird or a bat.

Figure 2.4. Pterosaur skeletons posed in flight. "Replica of *Geosternbergia sternbergi* skeletons": female (left) and male (right). Creative Commons image; source Kenn Chaplin from Toronto, Canada.

Lessons from two extinct giant fliers

Maybe it's impossible to make a real flying dragon?

We don't think so.

Why are we confident it'll work? As mentioned earlier and in the previous chapter, we know that dragon-sized, winged and likely flying creatures are possible because of our favorite such creature, Quetzalcoatlus (Figure 2.1). Quetzalcoatlus belonged to the Azhdarchidae group of Pteranodons, at least some of which almost certainly flew.

The two, potentially largest, flying creatures in the history of our planet have been Quetzalcoatlus (*Quetzalcoatlus northropi*) and its relative, Hatzegopteryx. Let's talk more about these two. As mentioned in the last chapter, Quetzalcoatlus was named after a feathered serpent god in the Americas (who could essentially have been a dragon) and Hatzegopteryx was named after an island that is thought to have once existed in the Cretaceous period, called Haţeg Island, in what is now Transylvania in Romania. We'll call these two amazing animals by the nicknames, Quetz and Hatz, for simplicity's sake from now on.

Given their average predicted wingspan of around 11 meters (and with maximum wingspans of about 16 meters, which is nearly 50 feet across), Quetz and Hatz must have been an amazing sight in flight, even if they didn't fly hundreds or thousands of miles at once. Just how big were these guys? For reference, their wingspans were larger than many houses are today, so they were dragon-sized and likely looked more than a bit like dragons too.

If only they were still alive now!

Funnily enough, some of these flying giants are thought to have lived in what we now call Transylvania, which tends to bring vampire bats to mind. Although we doubt that Quetz and Hatz drank blood (which would have been awesome!), the idea of a dinosaur-sized vampire bat or dragon is terrifying. Heck, even if it didn't drink blood, a carnivorous flying creature that is as big as a house would make a big impact. Although Quetz and Hatz had no teeth, we imagine that their sharp beaks could easily impale their prey. It's also possible some of their rather large prey were swallowed whole.

Given their weight (thought to be as much as approximately 500 pounds or about 225 kg), how did Quetz and Hatz fly? And are we sure that they did fly? Or could they have used their wings for other purposes like penguins, who use their wings as flippers to swim?

While there is still some debate about whether Quetz and Hatz could fly, since all we have to base our theories on is their fossilized bones, most scientists seem to think that pterosaurs, like Quetz and Hatz, were skilled fliers or that they could at least zoom around like gliders once airborne. We believe that Quetz and Hatz did fly, but that they also spent a good amount of time on the ground as terrestrial hunters, moving on all fours and perhaps soaring at times to capture certain delicious prey.

In fact, a team of researchers at a US research university called Caltech, led by aeronautical engineer Paul MacCready, created a Quetz model forty years ago that was reportedly able to fly. MacCready described the plan for the Quetz model prior to its creation in great detail.[ii]

Apparently, there was quite a bit of debate about this model's features prior to its construction, and there has been more debate about what the real Quetz was like. We found this passage from MacCready's plan, which is about people's assumptions about the biology of flight prior to the discovery of Quetz, to be encouraging from our dragon-building perspective:

> "Before the giant pterosaur's discovery, the size limits for biological flight were assumed to be much lower than an 11-meter-span flier. But nature is no respecter of performance limits assumed by man for biological creatures."

We won't put low limits on our dragon either.

[ii] http://calteches.library.caltech.edu/596/2/MacCready.pdf

Quetz and Hatz might be most closely related to today's birds rather than to the reptiles of our present-day world, such as the Draco lizard or Komodo dragon. Could we potentially give our dragon-to-be flight by leveraging the relatively close relationship of Quetz and Hatz to birds?

How do birds fly, and how does their flight compare to that of bats and insects? We always just assumed that these animals flapped their wings to create lift, or that they somehow soared like a kite, but there's a lot more to it than that. What exactly even is a wing and why does it evolve in a certain way to generate flight?

Figure 2.5. Bat wings mostly consist of thin bones and connected skin flaps called patagia. Image from Shutterstock.

Patagia for flight

Vertebrates that can fly possess a unique structure called a patagium (patagia when plural), which in general are flaps of skin that catch air. Patagia are universally shared by most flying vertebrates but are usually not found in adult terrestrial creatures except those that use their feet to paddle in the water. In vertebrate winged creatures, like birds, the patagium is more specifically the skin-based membrane that can be found between the extended digits in each wing, and it can be covered with feathers. As such, these structures are more easily seen in the wings of bats because they aren't covered up by feathers (Figure 2.5).

Figure 2.6. Note the prominent patagia in this flying squirrel. Image from Shutterstock.

Even gliding and soaring animals, like Draco lizards and flying squirrels, have patagia (see Figure 2.6). Note that the scientific name for flying squirrels is *Pteromyini*, which shares the prefix "ptero," meaning flight, with pterosaurs. (A funny side note is that researchers reported in 2019 that some flying squirrels appear to glow pink in the dark.[iii] Imagine glow-in-the-dark dragons!) And pterosaurs also had patagia that made their flight possible.

[iii] https://www.nytimes.com/2019/02/01/science/pink-squirrels-glow.html

What this all means is that our dragon definitely needs prominent patagia. Therefore, we either need to start our dragon with a creature – like a bat, bird, or Draco lizard – that already grows patagia during its natural development (and keeps them in adulthood), or we have to find a way to generate new patagia in a starter animal that doesn't already have them naturally. This possibility, of generating patagia, might not be as difficult as it first seems, as even humans have, what could be considered rudimentary patagia during our fetal development.

What do we mean?

Well, a patagium is essentially a web of skin (or membrane) that grows between digits. And as human fetal hands normally develop during a pregnancy, they are initially very much webbed in this way. This webbing, which for each space between digits is like a patagium, should disappear before birth, but rarely it is still there in newborn humans, due to glitches in their development. Notably, almost all vertebrate animals, at certain points during their fetal development, have webbing that could become patagia, including us humans as well. For example, in humans, as we've just mentioned, the cells that make up the webbing are mostly supposed to undergo a programmed type of cell death, called apoptosis, before birth. This also happens in other animals that aren't supposed to have webbing after birth.

Usually in animals, like us, that lack a lot or any webbing, the cells in the webbing die as instructed, but sometimes not quite in the way they are supposed to. Because of this, some people can end up a bit "webbed," which is a rare problem but one that can easily be fixed by surgery to remove the webbing. (Actually, most of us normally have a bit of flappy skin between our thumbs and pointer fingers (feel that area right now as you read this), which with some imagination could be considered a patagium.)

Then there's the problem called syndactyly, which is quite a feat to say and which causes one or more of an animal's (or human's) fingers or toes to fuse together. In less severe forms of syndactyly (where only some fingers or toes fuse together), excessive webbing can still remain between the unfused digits. From a dragon-making perspective, we could accentuate such webbing to form patagia by inhibiting apoptosis and, in parallel, we could extend the "finger" bones to form wings.

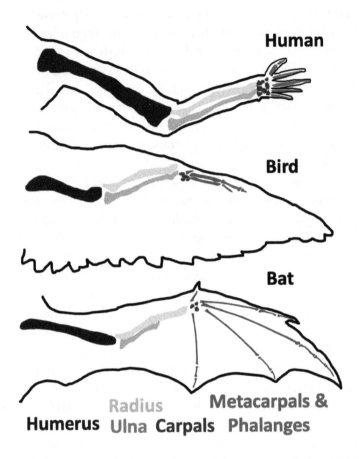

Figure 2.7. A sketch comparing human, bird, and bat arm/wing bones, highlighting similarities, in part based on [1] and inspired by a sketch at "Ask a Biologist" by Arizona State University.[iv]

Then there's the problem called syndactyly, which is quite a feat to say and which causes one or more of an animal's (or human's) fingers or toes to fuse together. In less severe forms of syndactyly (where only some fingers or toes fuse together), excessive webbing can still remain between the unfused digits. From a dragon-making perspective, we could accentuate such webbing to form patagia by inhibiting apoptosis and, in parallel, we could extend the "finger" bones to form wings.

But how could we stop apoptosis – this programmed form of cell death? Well, there are chemicals that can inhibit apoptosis, as well as genetic ways to block this process, which we could use to give our dragon patagia if we chose to start our dragon with a non-flying animal.

Some animals, such as ducks and other birds, have evolved to keep their foot webbing, even as adults. In these animals, the normally extensive apoptosis doesn't occur because the webbing is an evolutionary bonus that is important for their survival. It helps the foot of chicks and adults to have unique functions, such as being able to paddle more forcefully in the water. Since many birds and some reptiles already have some degree of webbing, one approach to giving our dragon patagia would be to retain or expand the webbing that our starting animal would naturally have as it develops.

It's not enough just to have patagia though. A flying vertebrate also needs to have specific bones in its forearms and "hands" – bones that are long enough to generate a flap that is itself substantial enough for the patagia to generate sufficient thrust and lift to get off the ground. We put "hands" in quotes because in flying animals the hand bones are often part of the wings. One could increase the forearm's and hand's growth during development, to match the greatly extended arm and finger digits that will form the scaffold of our dragon's wings. There are several families of genes involved in wing development, some of which are conserved among a wide range of animals, from insects (like the fruit fly) to bats and birds. (By the way, "conserved" means the sometimes extremely different organisms share the same gene, usually just with some small DNA changes.)

In fact, some of these same genes are involved in growing our own arms and legs but in flying creatures they have certain unique and special functions (Figure. 2.7). For example, certain genes can drive specific bones in the forearms, especially in the "hands," to grow differently to help bring about flight and to be tapered to reduce body mass. This unique gene activity in flying creatures is characterized by some of these genes being highly active (we sometimes refer to this as genes being 'switched on' or as genes being 'expressed'). The pattern of these genes' activity is also distinct in flying animals – we see them being switched on at certain times and in specific places as the forearm and digits develop.

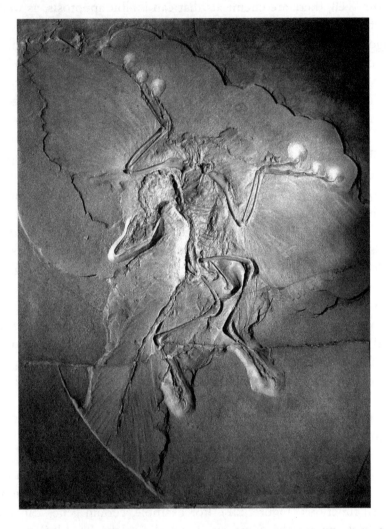

Figure 2.8. Photo of a unique specimen of *Archaeopteryx lithographica* fossil displayed at the Museum für Naturkunde in Berlin, Germany, by H. Raab, used under Creative Commons license.

There is one gene in particular, curiously called Hox-D11, that controls where and when some limb-patterning genes are turned on. For example, in humans and other animals, Hox-D11 regulates bone formation during digit development. However, where it is activated in birds is different from where it is activated in non-flying animals, such as in developing alligators and mice [2] or people. Together this means that if we can adjust the levels of genes like Hox-D11 up or down in certain ways, we may be able to give wings to previous non-flying creatures.

Getting a lift

Getting back to our question of how birds fly, birds have other unique features – in addition to patagia – that are important for flight, including powerful pectoral muscles, low overall body mass, hollow and lightweight bones, and various kinds of feathers. Birds use their strong pectoral muscles to flap their wings, and the wings – together with the patagia covered in feathers – create thrust. Through this process, birds can become airborne and, once in the air, they can flap or soar depending on the conditions.

During wing evolution – from dinosaurs to birds – certain wing shapes have been retained in largely similar forms for millions of years.[v] In birds, for example, the shape of the wing has been evolutionarily designed to move air both over and under it to create lift. Airplane wings bring about flight using a similar principle. As anyone who has ever made a paper airplane knows, the shape and design of the plane and its wings are tricky to get right – get it wrong and your plane swerves and crashes. What this means is that our dragon wings need to fit a specific design to work for flight.

If we start our dragon building with a bird, the wings won't need much changing other than being made larger. But if we begin with a non-flying creature like a lizard, how will we get the design right?

---

[v] https://evolution.berkeley.edu/evolibrary/article/evograms_06

Potentially, we could use artificial intelligence (AI) to help us to design the best dragon wings, in the same way that engineers use AI to help them to design new planes.

Planes and flying organisms have a couple of things in common. Lift is one of those shared attributes.

What is lift exactly?

The concept of lift is a surprisingly controversial and relatively difficult theory to study and explain.

We learned a great deal from reading this NASA page[vi] on lift, but in the end, the astonishing conclusion is that even the folks who put people on the Moon don't definitely know for sure how lift works.[vii] Or at least they cannot clearly explain it clearly to us.

Bird options

Let's say, for the moment, that we choose to use birds as our starter creature, then we still have a lot of choices to make. What or how many bird species should we use? A recent study roughly doubled the estimated number of bird species worldwide to nearly 20,000 [3]. A huge number of these predicted species don't even have names and there's no information about them. In that context, it's hard to begin choosing what bird to use to make our dragon, but we have some general ideas.

Since we want our dragon to fly, we need a flying bird. Sorry penguins! You are super amazing and cute, but not destined to be the basis of dragon building. It is kind of amazing though to imagine a dragon-like penguin sliding around on the ice or zooming through the seas.

This flight requirement also means that ostriches and emus are out of luck too. We recently saw a nature show about cheetahs hunting ostriches that ended badly for one particular ostrich. Imagine if that ostrich could have unexpectedly taken to the air or turned on the cheetah and let loose a huge breath of fire!

[vi] https://www.grc.nasa.gov/www/k-12/airplane/lift1.html

We'd also need a fairly sizable bird since we want to achieve a dragon-like size. This means that hummingbirds are out. Although, we can't help imagining a fire-breathing hummingbird igniting a feeder with a burst of flames if the food provided isn't quite to its liking. Still, even if hummingbirds are out of our project, we recently saw a piece in the New York Times (including a video) about how "cute" hummingbirds are also impressively fierce warriors, especially when competing with each other.[viii] Their need for battle may have even shaped their beak development.

Birds of prey like condors or other large birds like albatrosses (wingspans up to 11 feet or around 3.3 meters) could be good options based on their size, but we wouldn't want to use an endangered bird species here.

If we did decide to start with some kind of bird, we'd most likely want a species that gives us a large size, strong flying skills, and the ability to breed rapidly.

Birds or dragons of a feather

Would feathers help our dragon get airborne?

Birds have feathers and some ancient, bird-like dinosaurs, such as *Archaeopteryx lithographica*, had feathers too (see Figure 2.8, for a cool fossil with feathered wings). Bird feathers do help to make flight possible, but feathers are not absolutely necessary for flight (think about how bats, dragonflies, and other non-feathered animals are such elegant fliers).

Although some bird ancestors, including dinosaurs, had feathers, it is thought that pterosaurs most likely did not have true feathers. They are thought to have been sort of fuzzy as they had structures – called pycnofibers – that might have provided a layer of fuzz on their skin. A recent research article argued that pycnofibers had some

---

viii https://www.nytimes.com/video/science/100000006321699/how-the-hummingbird-bill-evolved-for-battle.html

notable similarities to feathers and may have influenced pterosaur aerodynamics.[ix]

We could give our dragon feathers for fun though or something like pycnofibers. From a utilitarian perspective, feathers, if designed properly, would almost certainly help our dragon to be an adept flier, so we have decided that our dragon likely should have feathers. And, yes, we admit it. Feathers could also make our dragon look like a badass, although if we don't get it right, we risk it looking more "derpy" (that is, goofy, sometimes in an endearing way) than frightening.

But what exactly are feathers?

In a sense, we can think of feathers as a sort of cross between hair and skin. Keep in mind that hair, and even fingernails, are essentially different types and extensions of skin too. Surprisingly, at a structural level hair is also highly related to scales such as on lizards. At the molecular level, feathers are made up of a number of different things, but primarily they consist of a protein called keratin, which is the most common protein in our own skin and hair too.[x]

A description of feathers by the Audubon Society[xi] captures their amazing, intricate complexities:

> "Think of a feather as a treelike structure. The trunk: a hollow central shaft, which ornithologists call a rachis. The rachis sprouts numerous branches, called barbs. In many feathers, such as those that form the shape of the wings and tail, the barbs are then further subdivided into twigs, so to speak, called barbules. On flight feathers, the barbs all grow in the same plane, like an espaliered fruit tree tacked to a sunny wall. The barbules of adjoining barbs, meanwhile, hook closely together to form a smooth and remarkably stiff surface that's critical to maintaining a durable yet streamlined aerodynamic form. On down feathers, by contrast, the

---

[ix] https://www.nature.com/articles/s41559-018-0728-7

[x] https://www.sciencelearn.org.nz/resources/308-feathers-and-flight

[xi] https://www.audubon.org/news/the-science-feathers

barbs twist willy-nilly in an ordered chaos that traps air and provides superb insulation."

Skin, nails, scales, and feathers all share a common cell type, called keratinocytes. The suffix "cyte" means cell, so these are literally "keratin cells." Millions of keratinocytes together form sheets of cells, and the arrangement of these sheets in part dictates whether a developing tissue turns into skin, feathers or hairs. Like hairs, feathers also have follicles that are responsible for making replacement feathers when feathers fall out [4].

Birds were not the first creatures to have feathers. The first known feathered creatures were a type of dinosaur called theropods (a group including *Tyrannosaurus rex*, although it isn't clear if *T. rexes* themselves had feathers), which did not fly and instead likely used their feathers for insulation, and perhaps for other things as well, such as for courting rituals.

In the book *The Tangled Bank*, by science writer Carl Zimmer, there is a great section on the evolution of feathers and flight [5]. Carl notes the remarkable coincidence that just one year after Darwin published his landmark work *"On the Origin of Species,"* a fossilized *Archaeopteryx* skeleton was found in 1860 in Germany (Figure 2.8). This fossil greatly puzzled those who examined it, since it had features of both birds and reptiles; in time though, this fossil provided biologists with new insights into how extinct dinosaurs and birds are evolutionarily related to the living birds we see today.

Archaeopteryx may in a sense have provided a link between birds and dinosaurs and might have provided a sort of evolutionary stepping stone between them. Did these winged, feathered beasts actually fly? A variety of research findings, including a report published in 2018, supports the idea of *Archaeopteryx* actually flying [6], which is encouraging for us on the dragon-building front since we need our dragon to fly.

As we know from birds, feathers have many other uses, such as insulation against the cold and for courting displays, which were likely put to use by feathered dinosaurs as well. If our dragon was feathered, it could use its feathers for functions other than flying too.

It may seem ironic that although dinosaurs were the first feathered creatures on this earth, they mostly didn't use their feathers to fly, while Pteranodons (which technically aren't dinosaurs) were likely to have been adept, but technically featherless fliers. On this, Carl Zimmer writes:

> "Because these feathered dinosaurs are the closest known relatives of birds, scientists can study them to make hypotheses about how flight evolved…It's possible that some feathered theropods evolved a flight stroke even before they could fly, to help them run from predators or capture prey. And in one lineage of small, feathered dinosaurs, this speed-boosting flapping evolved into true flight."

We found it cool that wings and feathers may have evolved before flight to serve other functions.

Genes and feathered flight

If we want to go the route of having both a winged and feathered dragon, toward those ends there are specific changes in gene activity ('expression') that we could try to induce in our developing dragon, potentially by using CRISPR gene-editing technology. A research team, led by a biologist called CM Chuong, has discovered, specific genes that are required for converting a skin scale into a feather [7]. These growth-inducing molecules and feather-making genes have funny names including Sox2, Zic1, Grem1, Spry2, and Sox18. According to this research, these molecules can instruct cells to form what are called "filamentous appendages" (long, thin tubular structures that stick out) that have been seen in feathered dinosaur fossils. These filaments may be like tiny baby feathers. The researchers propose that these genes may have had key roles in these kinds of steps in feather evolution.

There has been a lot of speculation about how pterosaurs evolved the ability to fly and, indeed, how they came to be the first flying vertebrate animal. Again, it seems likely to come down to how specific genes changed their patterns of activity to generate – from existing tissues, bones and muscles – the various larger body parts that are needed for flight [8].

Figure 2.9. Wingless (top) and normal winged (bottom) *Drosophila hydei* flies for comparison. Composite image of two Creative Commons images available for reuse and both credited to Brian Gratwicke.

What genes did extinct dinosaurs have?

Unfortunately, we're unlikely to ever know. Although Jurassic Park (and even some scientific publications as well, e.g. [9]) claimed that it's possible to extract dinosaur DNA from fossils, in reality, it's just not possible. One cannot retrieve, study or use dinosaur DNA, not even in tiny fragments. This is because dinosaur DNA has long since disintegrated [10]. Although DNA has a very strong backbone, even it totally falls apart over hundreds-of-millions of years old.

While we can study some less ancient DNA, it is basically impossible to study or to use dinosaur DNA. The only caveat here is that some amazing technical innovation in the future could make it possible to read fossilized dinosaur DNA or there might be some unknown reservoir of more intact dinosaur DNA.

On the other hand, some would say that the genomes of birds are likely to provide useful information about dinosaurs too, while others argue further that bird DNA *is* essentially dinosaur DNA because birds are modern dinosaurs.

Going back further in time, it is thought that the first flying creatures on our planet were insects. By flying in this case we mean creatures intentionally flying around with wings rather than just gliding or being randomly blown by the wind. You can imagine that the non-flying ancestors of insects were limited by being earthbound – by evolving flight with wings, their descendants opened up a whole new world of possibilities. Scientists have been able to give insects extra sets of wings by tinkering with the activation of certain genes[xii]. So, it might be possible to give our dragon entirely new wings, at least hypothetically.

Back in the 1980s and 1990s, biologists working on a species of fruit fly identified a range of genes involved in wing development. (Note that you probably have seen these little harmless, but sometimes annoying, tiny flies hovering over your bananas or other fruit at home.

[xii] https://www.nytimes.com/2018/03/26/science/insect-wing-evolution.html

They are nonetheless quite useful for research.) In particular, the one fruit fly species called *Drosophila melanogaster* is used the most in research labs. By physically removing some of these wing genes (in a process called "knocking out"), they managed to mess up or to entirely disrupt flight – they even created flies with no wings. It sounds a bit mean but by doing this, these scientists were able to discover the key factors involved in insect flight.

One gene found to be crucial for flight came to be called *wingless* because, when it was knocked out, the resulting flies had no real wings. (See Figure 2.9 for an image of a wingless version of another type of fruit fly called *Drosophila hydei*.) And from this came the discovery of other *wingless*-like genes in insects and in many other animals too (even in animals like us that can't fly). These *wingless*-like genes are all highly similar to each other, creating what we call a 'gene family'. In humans and in other animals that naturally lack wings, these genes were given the name *Wnt*.

It turns out that *Wnt* genes are crucial for many aspects of animal development (they don't just make wings) [11]. This means that altering them can have serious consequences – it can badly mess up the development of an animal, and can lead to other problems as well, such as cancer, especially if *Wnt* is too active. So, if we were to change these genes in some way, for example by using CRISPR, we would need to be extra careful not to create unwanted effects.

In the 2018 action-adventure movie *Rampage*, a CRISPR accident resulted in the creation of monsters, including a wolf with patagia that could fly or at least soar. At the time it was still being shown in many theaters, I (Paul) wrote about the depiction of CRISPR in Rampage as being somewhat ridiculous[xiii]. However, in theory, a well-thought-out project that used CRISPR to modify, for example, lizard embryos could hypothetically create lizards with patagia that may be could fly.

But wait – why would non-winged animals have the same genes that create wings in other creatures? The answer to this is that nature uses genes to regulate a variety of processes. As such, one gene (or gene family) can control or influence many processes. Evolution is very

---

[xiii] https://www.statnews.com/2018/08/15/movies-that-got-science-wrong/

efficient that way. *Wnt* genes across species are not so much directly "flight genes" or "wing genes," but rather, in a broader sense, they help animals to develop certain body regions with the right identity and function. Still, Wnt family genes, as well as the other genes we've mentioned, could help us to generate a dragon with wings that maybe can fly.

Dragon size

In a broader sense beyond just flight considerations, how big do we want our dragon to be?

Since our goal is to make a dragon that sparks an extreme "Dragon! Run!" kind of reaction in people, we figure our dragon should be about the size of a Pteranodon, which is around the size of a small putt-putt plane. Its weight would be much lower than you might anticipate, based on its dimensions, such as in the lower hundreds of pounds. However, in mythology and even in contemporary dramas, dragons have varied in size – from being enormous to being the size of a small cat or human (see for example the small dragon being slain by St. George in the painting shown in Figure 1.2 of Chapter 1).

We could, in theory, embrace trying to make smaller versions of our dragon – such a switch in our plan might be practically useful or even essential, given the laws of physics and the biological constraints on flight. In fact, we could make micro dragons that are about the same size as small birds. Our micro dragons could still be quite deadly – they could act in a coordinated fashion, almost like a fleet of drones, breathing fire. Given the relationship between surface area and the mass of an animal (mass grows faster than surface area as any given model organism grows), small creatures tend to function better.

As the potential target size of our dragon gets bigger, it has less surface area to support its mass, and so more stresses on that mass, which will likely make it physically more vulnerable to attack and even to just the stresses of everyday life as a dragon such as flying. We are encouraged, however, by the reality of Pteranodons having existed and the fact that most scientists think they were skilled at flying.

Where to call home?

Big, flying dragons the size of Pteranodons, or larger, pose other more practical challenges. Perhaps one of our toughest challenges is to find a place for our dragon to call home, and a huge dragon needs a proportionately big home. While it's cool to picture our dragon having a remote, open-air eyrie (an eyrie is the nest or home of a large bird such as an eagle, often found way up off the ground) atop a mountain, that's not entirely practical. Just imagine – every time our dragon is way up there, and we are way down here, it would have to come and get us. Or we would have to somehow get way up to it, not an easy thing to do unless we routinely wear jetpacks or have a speedy ski-lift-like system or super-fast elevator. Also, even way up there in its eyrie, the dragon would be vulnerable to attack from various enemies or to being kidnapped by plane. Still, an eyrie would be amazing and could afford some protection to our dragon by its remoteness.

So, we won't entirely rule an eyrie out, but the dragon needs a mostly indoor home or at least something like a cave up on a mountain. Ideally, their home would be close to our home. Initially, while we are making our dragon and then raising and training it, we need to keep it hidden as best we can. Both its safety and the safety of others around it are a consideration. A very large warehouse or airplane hangar might do the trick, providing space for flying lessons and fire-breathing practice. Our dragon's home is going to need exceptionally good ventilation!

What could go wrong with dragon flight?

We've already mentioned quite a few problems that we might encounter when creating a dragon that can fly, but there are many other challenges related to dragon flight. First, our dragon could end up not being able to fly at all, which would be so disappointing, but it'd still be a terrific beast. A hypothetical lack of flight could be due to the dragon weighing too much, such that its wings cannot produce the required lift to fly. And we might overshoot on its size and make our dragon as big as some dragons that are in the movies and on TV. Were we to do this, probably no wings – no matter how big – could result in flight.

Alternatively, we might create a dragon with a reasonably sized body, such as one or two hundred pounds, but its wings might end up being too small or its muscles not powerful enough to fly. It's also possible our dragon could fly but never learn to land properly, leading to anything from moderate crash landings to the full-on "splat" problem.

Most other possible disasters here relate to us riding our dragon as it experiences various glitches in its attempts to fly, leading us to experience our own version of the "splat" problem (think of a person whose parachute fails to open).

Some of the largest flying creatures to have ever lived likely did not initiate flying spontaneously on level ground because they were too large and couldn't generate enough power.[xiv] Instead, it's more likely that they initiated flight by gliding, having launched themselves by running down a hill or off a cliff, much like human hang gliders do today.

Here is an interesting quote from the earlier referenced *Gizmodo* article, which talks about one of the largest flying birds to have ever lived called *P. sandersi* that had a wingspan of as much as 20 feet or more (around 6 meters):

> "*P. sandersi* covered the sky around 25 million to 28 million years ago and had paper-thin hollow bones, stumpy legs and huge wings — all indicators of flight. Researchers used a computer program to estimate big bird's flight and figured it was basically a giant living hang glider, capable of reaching speeds up to 40 mph."

It's also possible our dragon would have no trouble flying or landing, but might become mad at us or tired of us and so decide to

---

[xiv] https://gizmodo.com/worlds-largest-flying-bird-was-twice-the-size-of-todays-1601476721

drop us from a great height. In our early efforts at riding our dragon, and perhaps while teaching it to fly, we could wear a parachute as a precaution. But that might not actually end up helping us if, for example, the altitude is high enough for a fall to kill us but not high enough to leave enough time for our chute to open.

So, when it comes to flying, there are some risks for both the dragon and for us. But, we still think flying is necessary for a dragon, very cool, and actually achievable.

How we could cheat on dragon flight

We predict that we'll get our dragon to be a skilled flier, but until we try, we can't be sure. It might be necessary to cheat along the way. One possible cheat would be giving our dragon wing extensions to boost its ability to generate the needed lift to get off the ground. We might also cheat by attaching some kind of propulsion system, kind of like a jetpack, to the dragon if it can't generate the needed power itself by flapping. Giving our dragon a satellite-based GPS system for navigation (like those in many cars and on cell phones) might be considered a small cheat if it isn't able to navigate well on its own.

Flying on

One way or another, we'll get our dragon to be a good flier, ideally without cheating. It's exciting to just imagine a flying dragon or watch one in a movie, but seeing a real one would be amazing. Imagine what it would be like to fly on the back of a real dragon.

References

1. Dumont, E.R., Bone density and the lightweight skeletons of birds. *Proc Biol Sci* 2010. **277**(1691): 2193–2198.

2. Vargas, A.O., *et al.*, The evolution of HoxD-11 expression in the bird wing: insights from Alligator mississippiensis. *PLoS One* 2008. **3**(10): e3325.9

3. Barrowclough, G.F., *et al.*, How Many Kinds of Birds Are There and Why Does It Matter? *PLoS One* 2016. **11**(11): e0166307.

4. Yu, M., *et al.*, The biology of feather follicles. *Int J Dev Biol* 2004. **48**(2-3): 181–191.

5. Zimmer, C., *The Tangled Bank : An Introduction to Evolution.* Second edition. ed. 2014, Greenwood Village, Colorado: Roberts and Company.

6. Voeten, D., *et al.*, Wing bone geometry reveals active flight in Archaeopteryx. *Nat Commun* 2018. **9**(1): 923.

7. Wu, P., *et al.*, Multiple regulatory modules are required for scale-to-feather conversion. *Mol Biol Evol* 2018. **35**(2): 417–430.

8. Tokita, M., How the pterosaur got its wings. *Biol Rev Camb Philos Soc* 2015. **90**(4): 1163–1178.

9. Asara, J.M., *et al.*, Protein sequences from mastodon and Tyrannosaurus rex revealed by mass spectrometry. *Science* 2007. **316**(5822): 280–285.

10. Morell, V., Difficulties with dinosaur DNA. *Science* 1993. **261**(5118): 161.

11. Yang, Y., Wnts and wing: Wnt signaling in vertebrate limb development and musculoskeletal morphogenesis. *Birth Defects Res C Embryo Today* 2003. **69**(4): 305–317.

Chapter 3

Fire!

Kindling a fire-breathing dragon

If our dragon looks like a real dragon and is a graceful flier, based on our efforts in the last chapter, then we've already made good progress toward building a dragon. However, countless flying creatures already exist, as well as various animals that bear at least some resemblance to dragons. We've already discussed some of these animals such as Komodo dragons and bats. Therefore, even if we achieve the look and flight skills of a dragon in a new animal, have we actually done something that is historic?

Maybe so, but we want more!

If we stopped at this point, our partial "dragon" would still be an entirely new creature and that's exciting, but we still have much more to accomplish.

As amazing as it is, giving our dragon flight feels less daunting than the next item on our wish list, which as far as we know, no living thing has ever been able to do: breathe fire.

We humans have made pretend versions of fire-breathing dragons for thousands of years – sometimes quite impressively, such as those given a spark of life during Chinese New Year celebrations (Figure 3.1) – and we (human beings in general, not us your authors) have even pretended to breathe fire ourselves (Figure 3.2). But we challenge you to imagine a real, living creature that can breathe fire on demand. We're hoping that even this exercise of imagination will knock your socks off. And what if there was a fire-breathing dragon right in front of you for real? You could feel the heat of its breath (from a safe distance, of course).

Since there have never been real, fire-breathing creatures, how do we even begin to give our dragon fire? There's no obvious creature to use as a starting point in the living world (or even historically, in the last few hundred years or in the fossil record) from which to bioengineer a fire-breather. In fact, you might think that it is impossible for any creature to breathe fire.

Figure 3.1. A fire-breathing dragon created during a 2003 Chinese New Year celebration. Image credited to 山脉.Use *via* Creative Commons license.[i]

We admit it is a challenge, but not an impossible one. If you look at humans who breathe fire for a living, such as in circuses (see Figure 3.2), it reassuringly shows us that fire-breathing might be doable and not deadly to the fire-breathers themselves. Human fire-breathers might

---

[i] https://commons.wikimedia.org/wiki/File:Fire_Dragon_dance.jpg

even give us some useful hints for how to go about giving our dragon this ability.

Figure 3.2. Fire-breathing humans use specific strategies to ensure they don't get burned even while making impressive flames, which are useful when thinking of ways to make a dragon that breathes fire. Image from Shutterstock.

For inspiration, we also looked at what others had written about when describing a dragon that can breathe fire, like the dragon Smaug in *The Hobbit*, and how it might be possible. We want to thank the author, Kyle Hill, for a particularly useful *Scientific American* article that strongly influenced our thoughts on dragon fire and this chapter.[ii] However, we also sparked some bright ideas ourselves.

We'll tackle the fire-breathing challenge by breaking our effort down into individual steps.

[ii] https://blogs.scientificamerican.com/but-not-simpler/smaug-breathes-fire-like-a-bloated-bombardier-beetle-with-flinted-teeth/

Fueling the flames

The first thing we wondered was how to fuel our dragon's flames. How can our dragon create fire without being loaded down with bulky flammable material or constantly having to refuel?

It's not as though we can regularly shove kindling down the poor animal's throat, not without risk of losing an arm or even our lives. The idea of throat splinters sounds awful for our dragon too and the dragon can't store fuel in its wings like a jet airplane.

And it'd be annoying for the dragon, and for us if it had to make regular pit stops to tank up at the gas station, as we do with our cars.

"Fill her up!" we'd tell the gas station attendant, who was cowering behind the pumps, as we fly in on the dragon's back. Other cars already there zoom away and those pulling up steer clear as they see the dragon, smoke rings puffing out of its nose. The attendant manages to squeak out, "No smoking allowed."

No, this isn't going to work as an option, plus each "tank" would likely cost us thousands of dollars.

So, what should we do for our dragon's fuel?

Gas is probably the best answer!

No, not gasoline as in the stuff we put in our cars, but flammable gases of different kinds.

The next question, of course, is – how will gas get into our dragon in the first place? And also: how will it be a "renewable resource"? Our dragon should be able to breathe fire any time it wants to, and ideally, do so repeatedly if in battle or just cooking marshmallows for us around the campfire.

Methane comes to mind.

Methane is a highly flammable gas that is produced during the digestive process of many animals, including us humans, making it a logical and practical choice that can be produced naturally inside our dragon's body without too much trouble, bioengineering, or cyborg hacking.

But our dragon would need to be pretty gassy.

And that's okay because some creatures are much better at generating methane than others. For instance, cows. Although cows are about as far removed from a dragon as it's possible to imagine, we surprisingly might want to borrow some lessons from cows or, more specifically, from their methane-producing guts.

Our motto here might be "Rumen has it."

Cows first digest the tough fibers found in grass, and in other food, in their so-called rumens (the first part of their digestive tract) and then produce "rumen liquor" (what a name, right?). This process results in the expulsion of a great deal of methane. In fact, methane released by cows alone is thought to be a major contributor to greenhouse gas production and potentially to climate change.

This entirely natural way of producing methane during digestion is catalyzed by microbes in the cow's rumen. Unlike humans, cows have multiple stomachs, and the first stop for food is the rumen – think of it as the cow's "first stomach" (rumens are also found in other related, appropriately-named "ruminant" animals). This is where digestion begins, and where very tough food is broken down and fermented, a process that is assisted by a slew of various microbes.

We started to think that maybe our dragon should have some kind of rumen for fermentation, populated by friendly bacteria. Most animals, even those lacking rumens, have loads (sometimes hundreds) of unique kinds of microbes in their digestive systems that can potentially contribute to flammable gas production by breaking down food and causing its fermentation. (By the way, fermentation means the process by which organisms such as microbes change some chemicals into ethanol, the type of alcohol that is in beverages like wine. Not only is alcohol intoxicating, but also it is quite flammable.)

Weirdly, some animals, are thought to not have microbes in their guts, but this remains somewhat controversial. For instance, birds were once thought to entirely lack gut microbes, but recent studies suggest that they do have them [1]. Nevertheless, some of the most common herbivores on the planet – caterpillars and similar creatures – seem to truly lack gut microbes.[iii] Is it possible the microbes in insect guts just haven't been found yet?

And in humans, our gut microbes are thought to be essential for our health and not just for making us feel gassy! But we might never know if this is true because we can't experimentally test this idea. It would be both unethical and highly difficult to prevent a human baby from acquiring gut microbes from birth (and to do so could make the baby seriously ill). Such a germ-free human would have to literally stay in a plastic bubble their whole lives, much like the restrictions on people who have "bubble boy" disease (a kind of genetic disorder where people are born lacking certain kinds of immune cells).

However, these kinds of experiments have been conducted in other animals, including mice. It turns out that our dragon might need a whole host of gut microbes (which are collectively known to some as the "gut microbiome," although technically that term just refers to the genomes of these microbes) for both fire-breathing and for its general health. Most mice used in research (including in the Knoepfler Lab) are "hyper-hygienic," which means that they are unnaturally clean and lack a diverse gut microbiome. Wild mice aren't particularly clean, by comparison, and have a whole range of gut microbes, which apparently help to keep them in good health. In fact, even a limited gut microbiome might help lab mice be healthier too.[iv] It's worth noting though that not all lab mice exposed to outdoor microbes (particularly nasty ones called pathogens) can survive, but those that do can end up fitter and stronger. At least in some experiments.

Farts, burps, and other fuel sources

What exactly are farts and burps?

If you think about it, every time we eat or drink anything, there's a good chance we swallow some air as we do, and most people swallow air at other times too. All that air has to go somewhere,

---

[iii] https://www.nature.com/news/the-curious-case-of-the-caterpillar-s-missing-microbes-1.21955

[iv] http://www.sciencemag.org/news/2017/10/how-gut-bacteria-saved-dirty-mice-death

right? So, we either burp it up or it gets pushed down through our guts until, at last, we fart it out.

The same goes with all those bubbles in carbonated (fizzy) drinks. They've either got to go up (burp) or down (fart). Stomach problems can also cause us to burp excessively, and intestinal infections can cause repeated farting as well (as many of you who've had a stomach bug will no doubt know).

But let's get back to the focus of making fuel for the fire in our dragon's gut. Other substances are produced during digestion – besides methane – that could be useful for our dragon, including flammable or even highly dangerous gases, such as hydrogen sulfide, hydrogen gas, and oxygen, as well as other substances such as alcohols. These substances make farts a lot more interesting in terms of their chemical composition. If this sounds like a toxic mix, you're right. In fact, farts not only smell bad but are also often literally toxic. Maybe that's no surprise to you. So for a dragon to breathe fire, we need their burps to be more like farts in terms of what's in them.

How could we turn our dragon's farts into burps and have some measure of control over the process?

Whether in a rumen or in a plain old undivided stomach, tweaking our dragon's microbes could create enough flammable gases, or a combination of gases, to produce fire. In the same way that some scientists are using microbes to make biofuels, we figure that certain combinations of microbes could produce a collective microbial ecosystem that would supply our dragon with sustainable biofuel for its fire.

We could derpily refer to this particular fire-friendly combination of microbes as the "fireome," since adding "ome" to anything makes it sound hi-tech, although this "ome" naming trend seems to be getting way out of hand.[v] The next question then is – what specific microbes might form the fireome?

There's a type of microbe called archaea, and a group of archaea called "methanogens" that have prodigious methane-producing

---

[v] https://phylogenomics.blogspot.com/p/my-writings-on-badomics-words.html

capabilities. So we might want to have some methanogens inhabit our dragon.[vi]

Our dragon would either have to store its highly flammable, gaseous fuel separately from the rest of its digestive system or have the microbes involved in gas production somehow kept away from the flames that could burn back into the dragon's gut and kill them. (Only certain microbes are particularly resistant to heat. It's possible no bacteria can survive direct exposure to fire.)

Figure 3.3. Photo of the Hindenburg exploding into flames. Why the hydrogen gas inside this flying vessel – the largest ever created by humans – ignited remains unknown. Image from Shutterstock.

In the human gut, one particular methanogen called *Methanobrevibacter smithii* is thought to be the main producer of "gas,"

[vi] https://www.hindawi.com/journals/archaea/2010/945785/

when we eat foods like beans,[vii] but there is a whole treasure trove of other microorganisms (bacteria, fungi, and archaea) that could populate our dragon's guts and that might be needed to make enough methane (and other flammable gases) to fuel fire.

If we could get our dragon to generate enough hydrogen, then could its presence also help the dragon to fly? Hydrogen is lighter-than-air and so could effectively lower our dragon's body mass, as well as fuel its fire. Think helium balloon, but filled with its neighbor on the periodic table, hydrogen, instead. But this is not only an admittedly nutty idea (would a hydrogen-filled dragon actually be able to fly better?) but a super risky one as well.

Have you heard of the Hindenburg?

It was a giant, rigid, blimp-like flying machine (an airship, also known as a dirigible). The Hindenburg was constructed by the Germans in the early 20th Century. They filled it with enough hydrogen to make it float into the air, despite being quite large and weighed down by passengers, crew, and baggage. As an airship, it worked pretty well, until it was destroyed in a fiery, explosive disaster (Figure 3.3). If our dragon were to store up too much hydrogen, and that hydrogen somehow ignited on the inside of the animal, it could have a similarly fiery, explosive and gruesome Hindenburg-like ending (hopefully, not while we are riding on its back). So, yes, we need to be careful here with our fuel plan and also with our dragon's design, if it is to carry flammable gases (hydrogen or any other).

Where exactly would the dragon store all this gas?

As anyone who has had too much gas in their gut can attest, it can be very uncomfortable or even downright painful, even without an unplanned gaseous explosion. If our dragon had about the same body volume (not mass, which would be much lower) as a large cow, with a cow-like digestive microbiome, perhaps our fire-breather could produce a massive amount of methane and store it in some kind of specialized digestive organ that could hold somewhat compressed gas without causing it discomfort.

[vii] http://www.sci-news.com/medicine/article00968.html

Even an average ruminant animal, like a cow, can produce up to 500 liters of methane gas a day [2]. We predict that we could double or triple that amount *via* microbiome engineering in a dragon whose digestive system might be similar to, or even smaller, than a cow's. As explained in Chapter 2, we need to keep the total body mass of our dragon as low as possible in order for it to be an adept flier.

Another hurdle for us to overcome – theoretically at least – is that some of this flammable gas could come out of not just our dragon's mouth but its rear end as well! And this wouldn't be particularly useful for our dragon. Still, what we know from how cows deal with gas tells us that this might not be such a huge problem. A common misconception about cows is that they fart away most of the methane they produce when, in fact, they mostly burp it out. This quirk of nature is useful to us in our efforts to create a fire-breathing dragon. We predict that with some modifications, we could get our dragon to produce some methane-rich, burp-like emissions – a good starting point for breathing fire. In contrast, we humans rid ourselves of our digestive gases mostly by farting while our burps are more often caused by swallowed air.

If our dragon were accidentally to produce a truly extreme amount of gas, it would need some way to release some of it. Because if it didn't – if the gas could not exit from either end – another gruesome possibility is that a gas buildup could cause our dragon to explode. Maybe sometimes our dragon could just burp without breathing fire?

Incidentally, dinosaurs are thought to have been prodigious farters, whose collective gaseous emissions were proposed to have changed the global climate of the time.[viii] By contrast, we humans can fart on average (a mere!) 20 times a day.[ix]

But if – despite our best efforts – our dragon emits its gases mostly through its rear end (so is more of a farter than a burper), well then,

---

[viii] https://www.smithsonianmag.com/science-nature/media-blows-hot-air-about-dinosaur-flatulence-84170975/

[ix] https://www.npr.org/sections/thesalt/2014/04/28/306544406/got-gas-it-could-mean-you-ve-got-healthy-gut-microbes

we could end up with a fire-farting beast instead. And that could be extremely hilarious but not so practical – it could lead to some unexpected problems, such as a singed rear end for our dragon, who isn't likely to find that amusing at all. Farting out flames is also not going to be that effective as a weapon (and imagine the dragon in that context trying to aim…but more on that to follow).

Dragon moonshine?

Yes, there are other flammable substances besides methane and hydrogen gas. Our research led us to a strange human affliction that could prove to be useful in our search for alternative sources of dragon fuel. It is a medical condition called auto-brewery syndrome. Like almost all people, those with this condition have a common type of yeast (*Saccharomyces cerevisiae*) in their gut that produces some alcohol. But in these people this home-brewed (or to be more precise, digestively-brewed) ethanol, in turn, intoxicates the sufferer because they typically lack the ability to properly break down alcohol [3].

Some animals naturally produce more alcohol than others. For example, there is a rare fish that lives in a naturally hot pool, called Devil's Hole, in Nevada. Appropriately enough, this fish is called The Devil's Hole Pupfish, and it reportedly produces more than seven times the amount of alcohol than any other fish living in cold water.[x] The alcohol (in the form of ethanol) is a byproduct of the pupfish's unique metabolism, which has adapted to its hot and harsh home [4].

If our dragon's guts could produce enough alcohol *via* fermentation, and if its body didn't excrete it or break it down too quickly (our dragon might need to somehow store the alcohol away from a natural alcohol-degrading enzyme, called alcohol dehydrogenase, or be made to produce less of the enzyme), then

---

[x] https://www.sciencenewsforstudents.org/blog/technically-fiction/nature-shows-how-dragons-might-breathe-fire

alcohol could substitute for methane and hydrogen gases, or it could burn mixed with them to add further fuel to our dragon's fire.[xi]

How would this work? Imagine a burp from a cow that had just helped itself to the farmer's moonshine – a burp loaded with both alcohol and methane. Now imagine the farmer lighting a cigar near that cow's mouth. Things could get interesting.

Cheats for fueling fire

If despite our best efforts, we couldn't get our dragon to produce enough of its own fuel to breathe fire, we suppose we could cheat by giving it a flammable material to hold inside its mouth immediately before fire-breathing, kind of like a human fire-breather does when they put on a show (Figure 3.1). But in practice, how could this work? Could our dragon carry around a small flask of fuel in one hand, and take sips from it now and then when it wants to breathe fire? That wouldn't be ideal.

Alternatively, our dragon could lug around a much larger fuel source, like a fuel tank on its back or around its neck, but this seems decidedly uncool and impractical. There are probably other ways we could cheat to supply our dragon with fuel, but we'd rather not.

The dragon diet

Almost every aspect of our plan to make a dragon will be influenced by what it eats and how that food interacts with its digestive system. It's not just the microbes in the digestive system, but the food coming in as well, for instance, that will help our dragon to produce fuel and, in turn, fire.

Our dragon's diet will need to be of a certain quality and quantity to fuel its flames, but it also needs to eat in such a way that its other main qualities aren't affected. For instance, if our dragon eats too much

---

[xi] https://www.npr.org/sections/thesalt/2013/09/17/223345977/auto-brewery-syndrome-apparently-you-can-make-beer-in-your-gut

overall or eats foods that are too caloric, it might soon become too heavy to fly, not to mention this may end up slowing it down while it's on land. However, given how much energy our dragon will be using to fuel its fire and flight, a more likely problem is getting it enough calories. If our dragon isn't eating right and is using up large amounts of calories, it could end up a scrawny thing.

Our dragon is also going to be fairly smart, which usually necessitates having a fairly big, sugar-hungry brain (more about this can be found in Chapter 4). However, as we discussed in Chapter 2, we are encouraged by the fact that some smart and bright birds, such as ravens and crows, that are great fliers, nonetheless have relatively light-weight brains.

In mythology and literature, dragons tend to be depicted as being carnivores with prodigious appetites (gobbling up whole large beasts such as cows and sheep, as well as people). If our dragon is a strict carnivore, is that compatible both with producing fuel for fire and with its other dragon attributes? For instance, gas-producing ruminants are strictly vegetarian, and they consume large quantities of low-calorie foods like grass. The digestion of a plant-based diet seems to produce the most gas.

So could a meat eater crank out sufficient methane and other flammable gases needed to make fire?

We think so, but we might have to give our dragon an occasional pot of beans or instigate "salad Sundays" just to be safe. The safest way to go in terms of fire production and its overall health will be to train our dragon to be an omnivore.

Another issue here is what happens after our dragon digests all the food in its diet. Is it going to be raining big dragon poops down from the sky? It's annoying enough when a relatively small bird poop hits your car windshield or lands on your shoulder, but imagine a 10-pound dragon poop, perhaps with partially-digested bones in it! We can hope that the dragon will just go the bathroom when not flying or only out over the deserted countryside.

We have ignition (we hope)

Our plan so far is for our dragon to produce its very own flammable mixture of gases and perhaps alcohol in its guts to fuel its fire. But wait, how will it ignite this fuel? We don't want our dragon to be a chronic smoker just to spark its fire, and we cannot be regularly lighting matches and chucking them down its throat, as that's not very practical (or cool). And yes, it would likely be extremely dangerous for us as well.

So, we've had to brainstorm some other ideas. Since matches came to mind (even if we don't want to be throwing them into our dragon), we decided to research how these little sticks create fire, as we reasoned that such knowledge could prove to be useful. Here's what we learned. The top of a typical match is covered with an unusual chemical called red phosphorus. When friction is applied to red phosphorus, a chemical reaction occurs during which the red phosphorus chemically transforms into white phosphorus (P_4). White phosphorus then combusts in air – and this is why a match bursts into flames when you strike it.

Learning about this process ignited (!) some ideas for our fire-breathing, dragon-building adventure. For instance, we could somehow integrate red phosphorus into the dragon's teeth (maybe as fillings?) so that when it gnashes them together, it could spark up its fire, much like striking a match tip. Or we could coat the dragon's tongue with phosphorus so that when the dragon rubs its tongue against the rough roof of its mouth, heat is produced *via* friction to light its fire. This seems pretty clunky to us though.

So what other ideas are there for lighting our dragon's fire?

We could "feed" our dragon a steady supply of rocks that contain red phosphorus. The rocks could then somehow rub together as the dragon grinds them. This might not be as strange as it first sounds. Birds and other creatures have an organ called a gizzard in their stomachs that contains stones, which they use to grind their food. Perhaps our dragon could have a similar organ under its throat to grind red phosphorus-containing rocks together, to create white phosphorous to ignite its fire.

Here's an interesting fact about gizzards. The rocks found inside of gizzards that help to grind up food have their own name – they are called "gastroliths," literally meaning "stomach stones." It turns out that dinosaurs also used gastroliths to aid their digestion (see Figure 3.4). Perhaps then, it's not such a stretch of the imagination to plan for our dragon to have special rocks to help it spark and then breathe fire – maybe they could aid its digestion too.

Even just eating flint might do the trick– it could also be stored in something like a gizzard and ground up there or in the mouth when our dragon wants to breathe fire. Flint would also be safer for our dragon than white phosphorous, which is not only very unstable and pyrophoric (self-igniting), but also toxic to the liver and other organs.

Overall, we are leaning toward flint gastroliths as being more practical to use than phosphorus.

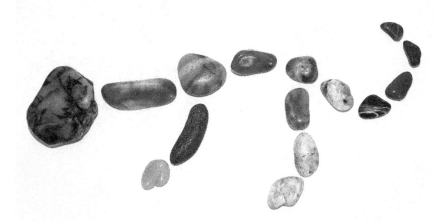

Figure 3.4. What are claimed to be dinosaur gastroliths arranged in a dinosaur-like form. Image from Shutterstock.

Imagine our dragon being able to store a bunch of flammable gases and alcohol, which it can eject from its mouth on command. In our plan, it would be able to grind flint against flint gastroliths – inside a gizzard-like organ near its rumen – to ignite this gaseous mixture on its way out. Wow, this could actually work! And with a little experimentation, these methods could be revised and perfected over time.

Some other pyrophoric compounds that could be useful to dragons, but which are also rare, include iridium, phosphine, and combinations of other things like iron and hydrogen sulfide.[xii]

Electrical ignition

We also have an entirely different way for our dragon to ignite its fire – one that is electrical rather than chemical. In this plan, our dragon could expel a nice concoction of flammable gases *via* a burp on command, and then ignite them electrically.

Figure. 3.5. An electric eel, whose special cells called electrocytes could be generated inside of and used by dragons to ignite their gaseous fuel to breathe fire. Image from Shutterstock.

[xii] https://www.sciencenewsforstudents.org/blog/technically-fiction/nature-shows-how-dragons-might-breathe-fire

While no creature on Earth that we know of has created fire, or even fiery sparks, using just its own physiology, there are creatures that can produce powerful electrical emissions. In fact, all animals have chemo-electrical activities in their bodies, such as in their nervous systems and muscles, but for most of us, the electrical current is weak even if quite important.

For instance, we humans have a nervous system that relies on electricity. Indeed, doctors sometimes need to measure the electrical activity of our muscles and brains, using tests such as an electrocardiogram (ECG, which measures the electrical activity of our hearts) or an electroencephalogram (EEG, which measures the electrical activity of our brains). Recent research has even suggested that information captured by an EEG might reflect the kinds of thoughts we are thinking, which itself is a rather disturbing thought in terms of privacy. Our nerves and muscles also interact with each other electrically.

However, some creatures have harnessed the electricity they produce (so-called bioelectricity) in much more dramatic fashion – in ways that we might want available for our dragon. Electric eels can produce and release a great deal of electricity. If our dragon could do the same, it could then generate the energy that would be needed to ignite the flammable gases it breathes out.

Electric eels (*Electrophorus electricus*, talk about a cool Latin name, see Figure 3.5) and other bioelectric creatures (such as electric rays) can produce and use electricity for a whole range of purposes, including to communicate with each other and to zap their prey. If we took a similar mechanism and placed it outside of a watery context, it could generate a spark strong enough to ignite our dragon's fuel.

So how do electric eels – and other bioelectric creatures, some of which use electricity only for sensory perception – make and release electricity?

These animals have special cells – called electrocytes – that work together to form electricity-producing organs (called electrical organs) inside their bodies [5]. Electrocytes, which are similar to both muscle and nerve cells, have special ionic properties that allow them to generate a surprisingly robust electrical discharge. These unusual cells are arranged in stacks in an animal's electrical organs, in much the same way that different chemical materials are organized inside batteries. In fact, pioneering electricity researchers studied electrical creatures like eels and took some valuable lessons from them.

We predict that we could engineer our dragon to have an electrical organ or two inside its mouth or throat (maybe one on either side) that could generate enough charge to spark its flammable gases to create fire. While most electric, sea-dwelling creatures have electrical organs along the sides of their bodies, some called Stargazers have their electric organs on their faces. This encourages us that it just might be possible for us to engineer our dragon to have its electrical organs inside its mouth or right on the edge of its mouth, to function as an electrical pilot light.

So the next question is – how would our dragon control its spark?

Again, we can look to the electric eel for ideas. It can stimulate its electrical organs to produce a coordinated burst of current when it senses and wants to zap its prey. With some training, our dragon could learn to release its electrical spark coordinately with breathing out flammable gases. Electrical animals can also control when and where they release their own electricity *via* what's called a pacemaker – a bundle of special cells that trigger other cells to release their electricity. Such pacemakers are not so different to the cardiac pacemakers that we all have in our hearts, which control our heartbeat – or to the artificial pacemakers that are implanted in people whose hearts don't beat properly.

Another cool idea inspired by electrical animals is the use of electrocytes and electrical organs as batteries to power cybernetic implants in humans if the right cells and organs could be bioengineered. In the same way, should we choose to engineer our dragon to have electrical organs to ignite its fire, we could use these same organs to power any cybernetic implants we might, in the future, add to our dragon to upgrade its capabilities.

For more on this idea, see Chapter 5 on potential upgrades and other features for our dragon. Overall, we think that electrical ignition might prove to be a more reliable way to generate repeated bouts of fire-breathing by our dragon, than grinding flint in a gizzard. But we might try both to see which works best in the real world.

Another idea based on electricity, especially if fire-breathing just doesn't work out, is for our dragon to "breathe" or rather shoot out electricity produced by its electrocytes as a weapon.

How we could cheat on ignition

Getting back to ignition of flammable gases, in the end, we could even cheat and implant a flint-based mechanical igniter in our dragon, maybe embedded in either a lip or tongue piercing? (Think about the mechanical igniter people use on their gas stoves or BBQ grills.) Alternatively, our dragon could just carry around a lighter in a little pouch or pocket (okay, we shouldn't even get started on what clothes our dragon might wear ... how even do you properly dress a dragon?)

Even if we don't want to cheat, our dragon might just be smart enough to think of a way to cheat if its own fire-lighting system wasn't working great one day. It could just find a lighter or a match, but that wouldn't be as fun, would it?

Protecting the dragon from its own fire

Remember the last time you ate piping hot pizza or some other food and burnt the roof of your mouth? We do.

This highlights another potential bump on the road to a living creature breathing fire, which is that it is somewhat dangerous to do. Actually, extremely dangerous, and far more so than eating hot pizza. How are we going to equip our dragon to be protected from its own flames?

Calcium silicate is a natural, fire-resistant substance found in limestone that we could use to line our dragon's mouth. Another possibility is for us to bioengineer some kind of fire-resistant mucus in its mouth. Once again, such protective methods would have to be tested out, but a combination of these approaches could work.

We also did some research on people who "breathe fire." We were wondering how on earth that's possible without burning yourself horribly and dying. According to Wiki How (you know, a very reliable source), the performer spits out a flammable substance (cornstarch or alcohol), which they then ignite using a lit torch. The flames are usually pointed up and away from the breather's mouth as a safety precaution.

Along these same lines, our dragon could "spit out" a flammable gas and ignite it just outside of its body (or right on the edge of its mouth) to prevent serious injury. This would create almost the same effect as the fire coming directly out of its mouth. Unfortunately, not only is this still going to be dangerous, but it will also take a great deal of practice. Just imagine our dragon experimenting with its fire spitting technique, getting it wrong, and an unfortunate mishap resulting that is deadly to onlookers. Our dragon could even die due to such a mistake.

Other tricks that human fire-breathers use include never breathing in while breathing fire, and angling their head upwards since heat rises. However, our dragon won't always be able to breathe fire upwards since its mouth will often be around the same height or higher than its prey.

We should probably consider fireproofing not only our entire research facility but also the outside of the dragon itself. If the dragon gets a little too energetic and by mistake blasts itself (for instance, its feet or tail), it would be catastrophic if the dragon itself isn't fireproof. We could coat the dragon in (the previously mentioned) fire-proof calcium silicate or mucus (more on mucus below), or perhaps its scales might be made to be thick enough to offer it enough protection.

We wondered if we might find inspiration for other possible fire-protecting strategies by studying different kinds of animals who can survive exposure to fire. The Pompeii worm was our first find. This worm uses filamentous bacteria as a fireproof barrier. Perhaps our dragon could as well. There could also be lessons to learn from those thermophiles (heat-resistant bacteria) that we mentioned earlier in this chapter.

Echidnas are little creatures that look like porcupines. They can withstand some fire by going into a state called torpor. Torpor is when an animal lowers its metabolism and – in the case of the echidna – hides in a thick layer of brush and shrubs. Apparently, this hibernation-like strategy helps the creature resist the extreme heat from the fire.

Overall, these strategies might help us to protect the outside of our dragon but also its insides as well.

Lessons from a beetle's butt

The Bombardier beetle is a funny creature.

It shoots hot acidic spray out of its butt.

How does it do this without seriously damaging both its butt and its overall survival? And what we can learn from it that might help our dragon to protect itself from fire?

We found that mucus was part of the solution.

And this makes sense because our own stomachs also use mucus to protect themselves from the "fire" of stomach acid.

What exactly is mucus, like that stuff that comes out of your nose? Sure, we all know it's gooey, slimy, and gross, but it's actually made of a special combination of proteins, salts, and in some cases germ-fighting substances. Without the normal mucus covering specific parts of our bodies such as our trachea (the living tubes we breathe through) and our guts, we'd quickly die. But would mucus protect against real fire? We can't be sure, but we'll give it a try and also develop multiple backup plans.

Of course, the Bombardier beetle's almost fiery rear-end emissions (Figure 3.6) are also interesting from the fire-breathing perspective as they provide an entirely different approach to engineering a dragon that can breathe fire or something like it. In fact, since these guys also fly, they are in some ways already, surprisingly close to dragons.

Figure 3.6. A Bombardier beetle spraying near boiling, acidic liquid. Copyright (1999) National Academy of Sciences, USA. Used with permission.

What can we learn from Bombardiers that would be useful? These little creatures are like the chemists of the bug world. If you ever took a chemistry class, or maybe you are a scientist yourself, you know not to mix or even to store certain chemicals near each other, as they can react in dangerous ways. Bombardiers both store and mix together on demand two reactive chemicals, called hydrogen peroxide and hydroquinone, inside their bodies.

Our old chemistry teachers would be amazed by this, but Bombardiers are able to pull this chemistry "experiment" off and in so doing often save their own lives by producing these geyser-like, near-boiling emissions from their butts and spraying them into the faces of their predators [6]. Bombardiers have even burned people.

Instead of shooting such a dangerous chemical reaction out of its butt, we want our dragon to breathe it out of its mouth. If our dragon were to use a Bombardier-like approach, we envision it storing a certain combination of dangerous chemicals separately like the Bombardier does in different chambers, or in its gullet or another organ. But we'd want our dragon's chemical mix to go beyond the

Bombardier's to include flammable chemicals that actually ignite on reaching a higher temperature.

Even though Bombardier beetles make and shoot out the chemical concoction from their rear end, they can aim it with great skill and can also spray it forwards, which has been the subject of some research [7]. We figure that our dragon could aim its fire coming out of its mouth, simply by moving its neck and controlling the intensity of the fire-stream.

If we just can't get our dragon to breathe actual fire, then another option is for it to spit out the same dangerous chemical reaction as the Bombardier. It would still be an effective weapon. Perhaps in the worst-case scenario our dragon could shoot the mixture out of its butt like the beetle instead of its mouth and still aim fairly well, but hopefully, it won't come to that.

Heat as a weapon

Another alternative to fire is for our dragon to breathe storms, as some mythological dragons were thought to do (as discussed in Chapter 1). Or it could shoot out just a wave of super-heated air. Certain animals and plants can generate a surprising amount of heat through the process of thermogenesis. If our dragon could do likewise – and build up enough heat inside its body – it could then breathe that heat out towards its target or prey to make it catch fire.

It's at least formally possible that our dragon could build up extreme heat in a particular part of its body, such as its rumen or lungs (if the lungs could be thermoprotected), and then breathe out that heat in a great burst. This could still be a powerful and destructive weapon even if technically the dragon isn't breathing fire.

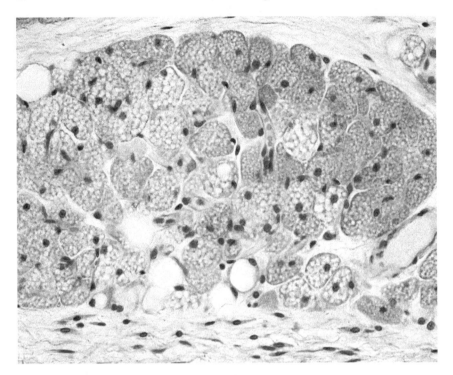

Figure 3.7. This image shows special fat cells (called adipocytes) that are part of brown fat, which forms lots of little droplets of fat in each fat cell. By comparison, white fat, which is the most common type in adults, usually forms one droplet per cell. Only brown fat can be used directly to generate heat. Image from Shutterstock.

Warm-blooded animals can regulate their body temperatures and produce heat. This includes producing fevers during illness and shivering to warm up, but there are other mechanisms of heat generation as well.

Brown fat, as shown in Figure 3.6, has a unique role in our metabolism, as it can be metabolized to produce heat – a process called thermogenesis. In humans, brown fat is thought to be present mostly when we are newborns in order to help regulate our body temperature.

Adults have mostly so-called white fat instead, so there is some interest in finding a way to convert this mostly white fat into what is thought of as healthier brown fat. Such a switch might help fewer people to become obese as well [8]. Or it could make adults sick for all we know.

Figure 3.8. A giant carrion flower *Amorphophallus titanium*. This plant generates heat to spread its smell, which mimics that of decaying animals to attract its prey including bugs. Image from Shutterstock.

People and birds both can maintain their body temperatures above that of surrounding air (which is called being "endothermic" or more commonly "Warm-blooded"). However, most animals and other organisms on Earth are cold-blooded including our potential dragon-starter animals the reptiles.

While young people (mostly infants) can use brown fat to help maintain their warmer-than-the-environment body temperature, to our knowledge birds do not have brown fat. Even so, many birds are surprisingly good at maintaining a consistent body temperature regardless of the coldness of the weather.

They use a combination of clever tricks to fight cold weather including fluffing their feathers, growing more feathers in winter, standing on one foot (remember their feet generally have no feathers to protect them), tucking their heads into their feathers, and hanging out together in protective groups like penguins do in the Antarctic.[xiii]

Unless we could introduce brown fat into lizards or birds, we might have a tough time building a heat wave-producing dragon from them. Scientists are studying how brown fat develops so it's possible that white fat might be turned into brown fat in the future as mentioned above for humans, but perhaps also in other animals.

While most plants have the same temperature as the surrounding environment, a few can generate heat, which they use for various purposes, such as to improve pollination, spread seeds, or to protect themselves from the cold.

But there are two specific examples of heat-generating plants that we'd like to share with you, as they are extremely unusual and quite impressive.

The dwarf mistletoe (*Arceuthobium americanum*) can heat itself up. Specifically, it heats its fruit, which then explode to scatter its seeds. While we've never witnessed this, the idea of a plant exploding part of itself to fire its seeds into the distance sounds extremely cool [9]. It makes good sense in terms of survival of the species too to get your seeds as far away from the parental plant as possible Maybe our dragon could use this thermogenic strategy to shoot shrapnel into its foes, in addition to breathing fire or heat.

The second example is equally impressive even if rather gross. The giant carrion flower (*Amorphophallus titanum*) smells like decaying meat, which is unusual enough. We've actually smelled one of these blossoms at UC Davis and it was like a dead body. The smell was incredibly powerful. It took a couple of hours to get that smell out of our noses completely.

The plant's appearance is almost prehistoric, or at least ideally suited for a dragon environment (see Figure 3.8) Out in the wild, this huge flower more effectively wafts that "deadly" smell around the jungle by using heat, to attract its prey [10]. Apparently, these prey include bugs that find the smell of death attractive because they eat

---

[xiii] https://www.audubon.org/how-do-birds-cope-cold-winter

decomposing tissue, so we don't feel so bad about the flower tricking and eating them.

If we applied this kind of idea to our dragon, perhaps we could create a beast that can kill its foes using hot, corpse-like bad breath as a weapon? This is so crazy it might actually work. It's also fun to consider other options for weaponizing the dragon's breath as well.

Nevertheless, we will focus primarily on fire-breathing as our dragon's weapon of choice.

What could go wrong and how might we die?

It's already abundantly clear that our attempts to give our dragon fire could go wrong in many ways, and it's easy to imagine how that could result in our deaths. For instance, our dragon could accidentally incinerate us with its bad aim while working out how to control the process. It could also accidentally burp out enough fire to cook us. It could set our house or research lab on fire by mistake with us inside. Or it could just intentionally decide to douse us in flames if we annoy it too much.

So the science and physiology of creating fire could easily backfire (ha ha) on us. As mentioned earlier in this chapter, the dragon could store up a lot of flammable gaseous fuel in its body and rather than breathing fire, it could accidentally ignite the gas to become a living bomb. If we happened to be with the dragon at that time, we'd almost certainly be goners. If we were a little further away, we might survive but be showered with half-cooked dragon parts.

In another unpleasant scenario, our dragon could accidentally (or on purpose, of course) release enough unignited gas to suffocate and kill us, particularly while indoors. We don't want it written on our tombstones, "Death by dragon fart" or "Killed by dragon burp," although they would be quite creative ways to die.

With these risks in mind, we've still got various ideas to try to keep things safe, or at least within the realm of not insanely dangerous. The spitting method might be the best and safest way for our dragon to create and aim its fire. We could set up some targets and have the dragon spit something harmless like, well, its spit (saliva) at them until its aim becomes accurate. If the spitting approach doesn't work out, our dragon could practice shooting out its flammable gas without actually lighting it and then, only later, practice with actual fire.

However, at the end of the day, we might have to make peace with the fact that building a living flame thrower is going to be inherently dangerous.

Take-home on fire

Overall, we are hopeful that with some practice we can get our dragon to breathe fire without harming itself, and we're hoping to stay uncooked ourselves as well.

In this chapter, we also talked about some alternatives to fire-breathing that could be equally amazing such as breathing storms or shooting electrical beams. In the end, our creation needs some devastating kind of weapon to really be a dragon.

References

1.   Videvall, E., *et al.*, Measuring the gut microbiome in birds: Comparison of faecal and cloacal sampling. *Mol Ecol Resour* 2018. **18**(3): 424–434.
2.   Pimentel, M., R. Mathur, and C. Chang, Gas and the microbiome. *Curr Gastroenterol Rep* 2013. **15**(12): 356.
3.   Hafez, E.M., *et al.*, Auto-brewery syndrome: Ethanol pseudo-toxicity in diabetic and hepatic patients. *Hum Exp Toxicol* 2017. **36**(5): 445–450.
4.   Heuton, M., *et al.*, Paradoxical anaerobism in desert pupfish. *J Exp Biol* 2015. **218**(Pt 23): 3739–3745.
5.   Markham, M.R., Electrocyte physiology: 50 years later. *J Exp Biol* 2013. **216**(Pt 13): 2451–2458.
6.   Aneshansley, D.J., *et al.*, Biochemistry at 100°C: Explosive secretory discharge of Bombardier beetles (Brachinus). *Science* 1969. **165**(3888): 61–63.

7.   Eisner, T. and D.J. Aneshansley, Spray aiming in the Bombardier beetle: photographic evidence. *Proc Natl Acad Sci USA* 1999. **96**(17): 9705–9709.

8.   Seale, P. and M.A. Lazar, Brown fat in humans: turning up the heat on obesity. *Diabetes* 2009. **58**(7): 1482–1484.

9.   deBruyn, R.A., *et al.*, Thermogenesis-triggered seed dispersal in dwarf mistletoe. *Nat Commun* 2015. **6**: 6262.

10.   Barthlott, W., *et al.*, A torch in the rain forest: thermogenesis of the Titan arum (*Amorphophallus titanum*). *Plant Biol (Stuttg)* 2009. **11**(4): 499–505.

Chapter 4

Dragons on the brain

Make up your mind

We admit it – we have dragons on the brain. We are obsessed with making a dragon!

And the nature of our dragon's brain is an area of great interest and excitement for us too, so much so that you could say we have "dragon brain on the brain."

Our dragon's brain will both keep its body running (you know, basic things like keeping its heart beating, lungs breathing, and so on) and run higher-level functions, like thinking out complex problems, talking and – if all our plans work out (see the previous chapters) – orchestrating its flight and fire-breathing too.

Given our borderline obsession with the brain of our future dragon, you can imagine that it has been kind of tempting to try to push things to the limit technologically. For instance, we could try to make our dragon as smart as possible, perhaps as intelligent as Albert Einstein or Marie Curie, or even smarter. However, not only is this technologically going to be nearly impossible, but also as we'll discuss later in this chapter, there are good reasons why trying to make a genius dragon would likely be a terrible idea. Just one of the many potential problems could be that our dragon comes to view us as beneath it – and then it might just blow us off and go elsewhere or even kill us.

On the other hand, if our dragon's brain ends up at the other end of the spectrum, such that it isn't nearly smart enough, we'll likely face other big problems as well. It could be untrainable. It could end up accidentally breathing fire on us or dropping us while flying because it is just too stupid to realize the consequences.

For these reasons, a certain level of intelligence is a major must-have for our dragon. Furthermore, when it comes to brains, it's not just about running the body and being smart, but also about the mind and personality, maybe some would even say the soul, of our dragon too. These are kind of crucial to not mess up, but they are fairly nebulous as well.

How much of a role does genetics play in mind and personality *versus* environment and upbringing? It's likely a combination of these influences. How does the structure of the human brain impact personality and identity? At this point, brain scientists are likely to say collectively, "Who knows?" or "We're making some progress on this question, but get back to us in ten years."

Also, what exactly is a brain?

It's not just the "seat of the soul," as epic as that sounds, but also and more practically, it is the computer in charge of making sure that our bodies (and those of all "brained" creatures) run properly.

As a side note, there are in fact some creatures that are technically animals without a brain, like the Scarecrow from The Wizard of Oz, who sang, "If I only had a brain." (In fact, singing itself involves some complex brain functions so the idea of a brainless Scarecrow singing is even more nonsensical than in might first seem!)

But anyway, what about those real creatures that lack brains? Well, certain sea creatures, like jellyfish, starfish, and sponges, are literally brainless. And strikingly, some researchers believe that some of these brainless animals, such as the sponges, did once have brains earlier on in their evolution. This means that at some point these creatures evolved away from having a brain, which in turn means that they found it more useful evolutionarily to not have one.[i] (Maybe earlier on in evolution when they still had brains they were singing, "If I only *didn't* have a brain?" Yes, they can't sing, but it's funny to picture that idea.)

---

[i] http://www.bbc.com/earth/story/20150424-animals-that-lost-their-brains

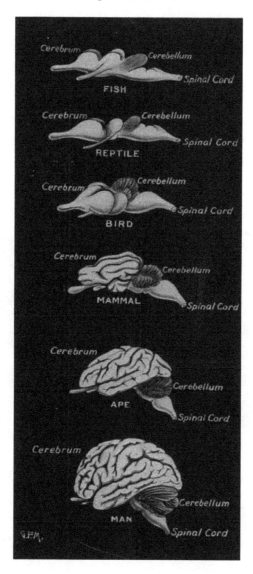

Figure 4.1. Illustrations of various animal brains, which increase in complexity from top to bottom. Public domain image from "*The Outline of Science*" by J. Arthur Thomson, 1922. Note, the brains of birds and reptiles are smooth and simple brains, which generally indicates lower cognitive function, as compared to other brains, like those of primates, which are larger and more folded.

However, brainless creatures are relatively rare. All vertebrate animals, meaning those with a spine, have brains, like us humans. In vertebrates, the basic structure (organization of the major parts) of the brain is highly conserved, which means that it has essentially remained unchanged in the big picture sense through evolution and so is shared by many animals (see Figure 4.1). Generally, in vertebrate brains, the cerebrum is up front and the cerebellum is at the back, together with a spinal cord, *via* which the brain communicates with the rest of the body. In the middle, not surprisingly, is something called the "midbrain."

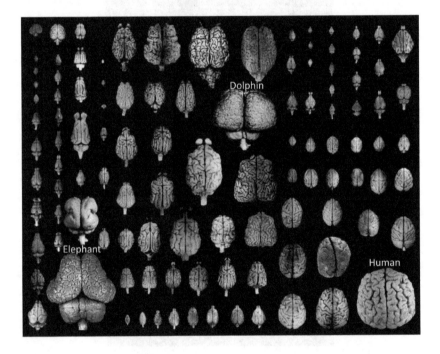

Figure 4.2. Illustration of various animal brains from *Sizing up the Soul's Seat*, by Florian Maderspacher.[ii] Note the rich diversity of brain sizes and shapes. Used under license by Rightslink. Brain images originate from The Brain Museum, reused with permission. The overall image was slightly modified to add the names of the animals having the three biggest brains above those brains: the elephant, human, and dolphin.

[ii] https://www.cell.com/current-biology/fulltext/S0960-9822(16)31151-4

Although brains come in all kinds of shapes and sizes (see Figure 4.2), the cerebrum (or cerebral cortex) is important in each animal for cognition. Other brain regions perform roughly equivalent functions in different animals as well. For instance, the cerebellum, which sits at the back of the brain, mediates coordination in all animals that have one. Human body movements are highly accurate and defined (at least most of the time, but in some of us more than others.) For this reason, tremors can be a hallmark of problems with the cerebellum.

Some also believe that the cerebellum plays important roles in cognition too. Yet brains are complex structures and are not entirely understood – as such, debates continue about how its anatomy correlates with its functions.

A dragon's brain

Since the brain runs the show, we need to pay close attention to what kind of brain our dragon grows. If we manage to make a dragon that physiologically can fly well and breathe fire, and that basically looks like and functions like a dragon, it would be a great accomplishment. But we aren't quite at the finish line yet.

It also needs a certain kind of brain or else it won't quite function like a dragon at all and might not survive too long either. And we might not survive too long as well.

What could go wrong with our dragon's brain?

Many things.

First of all, the brain controls the rest of the body. So, all the work we put into the wings for flight – and the biology behind fire-breathing – could easily be for nothing if our dragon doesn't have a healthy brain to run it all. The importance of the brain is highlighted by the fact that a person's body can be healthy, but if they are found to be "brain-dead" because their brain no longer works, then they are themselves considered to be dead. So in us humans and in any other creature with a brain, its function is essential. The same will be true for our dragon.

First, let's focus on getting the dragon's intelligence 'just right' because all kinds of bad things, including our own deaths (which is, we realize, a recurring theme in almost every chapter of this book), could result from our dragon being too dumb or too smart. For these reasons, we are aiming for a level of intelligence that is "just right" – not too smart nor too dumb (think Goldilocks and her favorite porridge). We believe that 'just right' intelligence for our dragon will provide the best results – for us, as its creators, and for the dragon's own happiness too…and maybe for the world as well, including you.

With this "just right" intelligence goal in mind, what kind of brain does our dragon need and how the heck are we going to give it to him or her? Animal brains vary so substantially in size and architecture (see Figure 4.2 again for a striking assembly of brains) that there are many options to choose from for our dragon, in theory at least.

In Figure 4.2, the three largest brains shown are those of an elephant, human, and dolphin, with the elephant brain being the largest. Yes, humans don't have the largest brains, and it's even possible that we aren't the smartest creatures on Earth. And are big brains important for intelligence? Or are they sometimes – in certain rare circumstances – a disadvantage for survival (kind of similar to a sponge as mentioned earlier)?

Even animals with relatively tiny brains get along just fine and, in a way, you could say that each animal has a brain that is just right for its own needs.

One other thing to notice from Figure 4.2 is how different brains look compared to each other in terms of their structure – how folded they are, their overall shape, and their differently sized brain regions, which vary in their relative proportions to each other (such as the cerebellum relative to the brain as a whole). These differences are generally attributed to the specific characteristics or needs of each animal and arise partly because of specific patterns of gene activity during embryonic and fetal development.

In Figure 4.3, you can see an example of a cerebellum – it's the more convoluted and wrinkly part at the back of the brain (in each view of the brain, the cerebellum is oriented toward the middle of the image). These brain pictures come from a type of bat, called a flying fox (*Pteropus giganteus*). (Note that the "Ptero" prefix refers to flight, as mentioned in the other chapters and as shared with other animals, like Pteranodons.)

Although this bat's scientific Latin name literally means flying giant, it has a relatively small brain – just about an inch long (and its body is very small too). So this brain size seems to suit this particular bat perfectly for flying and for echolocation, another of this bat's capabilities. Without any other changes to its body, suddenly having a brain twice the normal size wouldn't help this bat, but instead could kill it. Everything is relative.

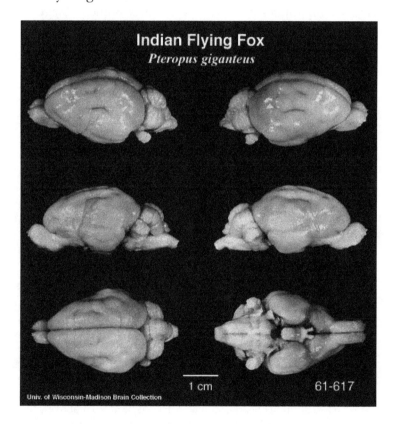

Figure 4.3. Bat brain from *Pteropus giganteus*. Brain image originates from The Brain Museum, reused with permission.

Similarly, for our dragon, we don't care so much about the exact size of its brain, but rather its functions, such as intelligence. While brain size matters, it does not always correlate with intelligence, which adds a complicating factor. As mentioned in earlier chapters, an extra-large brain would also be both an energy drain and more weight for our dragon to carry as it flies.

Beyond absolute brain size, the brain-to-body mass ratio (that is, how big your brain is relative to the size of your body) is probably more important as a determinant of intelligence. For example, the larger the animal, the larger the brain it typically has. This kind of biological axiom applies more broadly. For example, a bigger animal will generally have larger organs, with some notable exceptions – some dinosaurs are known to have had exceptionally large bodies but puny brains.

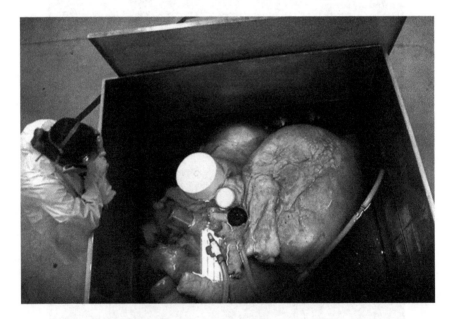

Figure 4.4. The enormous heart of a blue whale, as seen from above while being preserved by researchers who are smaller than it. Image source, Apeksha Roy. Creative commons license on picture available for re-use.

Still, this rule generally holds true. A blue whale, for example, has a heart that is much bigger than an entire human (see Figure 4.4), weighing in excess of 1,000 pounds (450 kg).[iii] But its heart isn't *better* than a human's (which weighs less than 1 pound or half a kilogram) or even than that of a mouse (which weighs less than 200 mg) just because · it's so big.

Even for human brains, bigger isn't necessarily better, as extra-large brains can be associated with various conditions. For instance, people with autism generally have unusually large brains and a greater cortical surface (which is otherwise generally linked to higher cognitive capabilities in non-autistic people) relative to their non-autistic siblings [1].

Although those with autism can have normal, or even higher than average intelligence, they often also have impaired cognition and other brain-related challenges. Thus, bigger isn't necessarily better when it comes to brains and that's probably true for dragons too.

Later in this chapter, we discuss the various brain features that might contribute in unique ways to intelligence, and how they factor into the design of our dragon's brain.

Picking an unpickled starter brain

When you buy your first computer, you have specific factors and features in mind, like the brand (Apple, Dell, etc.), its size (e.g. laptop or desktop), the amount of memory, and so forth. What key features would we look for in our dragon's brain and what type of brain would it be best to start with?

In his article *Sizing Up the Soul,* biologist Florian Maderspacher has this to say about brains, and in particular how it might be futile to try to pick brains that are the most "able," whatever that means:

[iii] https://www.whalefacts.org/blue-whale-heart/

> "If a pickled human brain is placed among the brains of other mammals, it doesn't jump out immediately. Not in a way that would adequately reflect our cognitive accomplishments. If an alien were to stroll through a museum of brains in vats and asked to pick the brain of the most intelligent species — the one that was capable of pickling all the other brains — which one would she pick?"[iv]

It's hard to choose a "best" brain in such a situation based on its appearance alone. And again, one wouldn't know, for example, what the overall body mass of its previous owner had been. Even with more information available to you, what actually constitutes a "best" brain?

And imagine if there was a strange but remarkable museum that exhibited the living, transplantable brains of creatures, being kept alive in vats of nutrient-rich broth. Which of these brains would we pick for a dragon and why? Yes, we realize this is starting to sound a bit Frankenstein-like (we could even send out our assistant "Igor" to collect a brain for us). But we need to think carefully about the options for our dragon's brain.

One option would be a bird brain, and we don't mean that in a bad way even though "bird brain" is often used as an insult to imply that someone is of low intelligence. In reality, there are some extremely smart birds. Studies of bird intelligence overall suggest that the term "bird brain" is often a misnomer. For instance, crows and ravens are highly intelligent. They have been known to use tools and to remember the faces of humans who have perhaps bugged or hurt them. They even sometimes seek revenge, which is why researchers who work with crows and ravens sometimes find themselves having to wear a mask to hide their identities from these smart birds, who may not appreciate being part of experiments.

What are the hallmarks of being an intelligent animal? Well, there are quite a few: having a good memory, being able to use tools, bonding with a mating partner, forming family groups, playing,

---

iv https://www.cell.com/current-biology/fulltext/S0960-9822(16)31151-4

recognizing one's reflection, and finally singing or having some other form of language. Birds "sing," and some of them use tools, which puts them in a relatively high-intelligence bracket. However, some birds are indeed not too bright. Also, the brains of some birds and other possible starter creatures such as lizards, have a relatively simple form (see Figure 4.1).

And what about using a lizard brain to build our dragon's brain from? We weren't able to find any evidence that certain lizards are smart in some way, even if they actually aren't as super dumb as many people think. So if we started with a lizard brain, we would need to focus on making it more advanced for our dragon.

The more we thought about it, the more we leaned toward using a bird brain as the basis for our dragon brain, but not just that of any bird. The smartest birds are thought to include the raven and others in its family, the Corvidae.[v] It's not entirely clear what makes them so smart, and smarter than most other birds, but their brain-to-body mass ratio is relatively high, which may play a role.

We envision our dragon brain being generally similar to the brain of a smart bird, but somewhat larger and perhaps with some more of those cortical folds that primates like us have that boost intelligence. We'd want more, but not too many of these folds (called gyri) as we don't want our dragon to be too smart, as that could cause us problems. Again, we'll be aiming for a sweet spot when it comes to brain growth. Not too big and not too small.

Then there's personality and other non-intelligence-related brain attributes. What if we achieve our ideal level of intelligence for our dragon, but it comes with a pathological personality? Or perhaps while not literally pathological, our dragon could end up being anything from a high-functioning sociopath to just extremely annoying?

What if our dragon is a pacifist to the point of not even wanting to act in a dragon-like manner at all? With that kind of mindset, it could end up not wanting to fly around terrorizing the world (not that we would want it to do that), or it might even do something as simple as accidentally scare a child and never want to breathe fire again. Ever!

---

[v] https://news.nationalgeographic.com/2017/07/ravens-problem-solving-smart-birds/

It's not that we want our dragon to kill people, but if it becomes the next Mother Teresa or Gandhi, then that might take some fun out of the whole adventure. Funnily enough, a recent study of Komodo dragons found that they tend to be extreme homebodies.[vi] What if our dragon just wants to spend all its time on the couch, watching TV or playing on its iPhone?

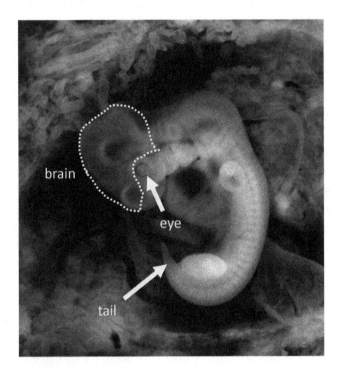

Figure 4.5. An extraordinary photo of a developing human embryo very early (7th week) during pregnancy that ended up outside the uterus, which makes pregnancy life-threatening to the mother. Image credit: Wikimedia open source image by Ed Uthman, M.D.[vii] The image has been modified to change it to black and white, and indications of specific structures including the brain and, yes, tail (recall that during development we humans do have tails for a while).

---

vi https://www.nytimes.com/2018/11/13/science/komodo-dragons.html

vii https://en.wikiversity.org/wiki/WikiJournal_of_Medicine/Tubal_pregnancy_with_embryo

For these kinds of reasons, engineering our dragon's brain is going to be difficult. How are we going to get everything right?

How to grow a brain

If you think about it, a simple, broader question pops up first.

How do any of us animals grow a brain?

The natural process of growing a brain of the right size and correct organization in normal animals – from worms, flies or humans – is absurdly tricky. It's a highly orchestrated, complex process, no matter how simple your brain is, during which any number of things can go wrong to cause neurological disorders and even death.

We also need our dragon to develop properly as an embryo and as a fetus because broader problems can mess up the developing brain and, of course, those broader problems might result in us not having any dragon born at all.

If our dragon has a mother (and what kind of creature would that mother be? We briefly discuss this in Chapter 6) and its mother had health problems during pregnancy, the developing brain of our growing dragon fetus could be affected in many ways, leading to problems such as microcephaly (which is when an animal has a small head and small brain, usually accompanied by impaired cognition).

In the real world, microcephaly can be caused by a mosquito that carries a particular virus – the Zika virus. When a Zika-carrying mosquito bites a pregnant woman, the resulting viral infection can sometimes interfere with her growing baby's brain development, to cause microcephaly. But microcephaly can also be caused by errors (known as mutations) in the DNA sequence of specific genes. As a result of all these potential issues, in our research we have to worry about both the health of our developing dragon and that of its surrogate mother too. It's all going to take more than prenatal classes and dragon vitamins.

Rather than engineering a dragon brain from scratch (for example, by growing brain-like structures in the lab, see more on "organoids" below), our plan is to modulate the brain growth and development of an existing creature (such as a lizard or bird), which we will use as a starting point to make our dragon. And we could start tweaking its brain development at various times, using different approaches.

We could do this in the embryo, or in the mother's egg, or using stem cells (these are special cells that can in certain cases form pretty much any cell type in the body when exposed to the right signals). In this way, we'll nudge the resulting living dragon to be somewhat smarter. This tweaking of brain development would be far easier than trying to make a brain from scratch.

We also don't envision brain transplants as being a realistic option. And this is despite an Italian doctor, Sergio Canavero, claiming to be on track to perform head transplants[viii] and possibly brain transplants in the future. (Head transplants have also been featured in various movies over the years, including in the 2017 film, *Get Out.*) It also wouldn't be particularly practical either to give our dragon a brain or head transplant, given that even the gung-ho Dr. Canavero estimates it will cost at least US$100 million to successfully accomplish.[ix]

Normally, brain growth is a highly complex and conserved process, which means that a lot of animals grow their brains in somewhat similar ways, even humans and fruit flies! And many of the same types of cells, regulatory factors, and structures are involved, and the same types of genes and proteins too.

So how do we animals normally grow our brains 'in utero' (that is, while we're developing inside our mothers)?

First, a little collection of cells is specified to eventually become the central nervous system, including the brain. In science, "specified" means telling cells what to become so that they eventually form the right tissue or organ in the right place. It's akin to young people going to a specific school to learn a trade like computer programming, engineering, or plumbing, which becomes their profession later on in life. There are a lot of molecules that control this vital and essential process for the brain. And what cell specification achieves is rather

---

viii https://www.popsci.com/first-head-transplant-human-surgery

ix https://www.usatoday.com/story/news/world/2017/11/17/italian-doctor-says-worlds-first-human-head-transplant-imminent/847288001/

amazing – it can help to program simple cells to create complex structures like the brain and nervous system.

The cells that will ultimately make the brain and spinal cord don't immediately look any different than the surrounding cells, which will become other things. But inside these specified "brain-to-be" cells, a unique program of gene activity is activated. Some brain-development genes are turned on a little, or they get ready to be turned on later in development. Other genes are switched off or turned down – these suppressed genes would otherwise tell cells to become other kinds of things, such as blood or bone or gut (you name it). During embryonic and fetal development, it's not enough to tell cells to make a specific organ, the developing body has to tell them not to make other stuff as well in a given region.

Going back to the human job training analogy again, it'd be like telling a young person they can only have one main profession, and can't try to be a doctor, painter, scientist, plumber, and computer coder all at once. In humans such parental advice doesn't always work, but in stem cells most of the time they end up just changing into the one "right" final type of cell they are "told" to be.

The "brain-to-be" cells are also growing and dividing very rapidly, sometimes doubling their numbers more than once a day. That may not sound fast, but remember that South Asian folk tale "One Grain of Rice" by Demi about the woman who asks for double the number of grains of rice every day. She started with just one grain, but eventually got billions of them. The same kind of deal happens with the cells that form the brain. Cell division quickly can quickly change one cell into billions.

At this point in early development, the "brain-to-be" cells (and those of the central nervous system) have some genes turned on (or poised to be turned on) that instruct them to become a brain cell. And they have other genes turned off – genes that would turn them into other things, like the heart or liver.

Despite all this activity, the brain and central nervous system are not much to look at early on in development; they also look pretty much the same across a wide variety of animals at this stage. In fact, early human embryos appear to be quite similar to the embryos of other animals at comparable stages of their development. There's no obvious outward sign of the powerhouse brain that is to come later.

As development continues, another program kicks in. As a result of this program, the "brain-to-be" cells start to make new cells that will form the mature brain, rather than just dividing to make more of themselves. Soon after, some immature neurons start to appear, and then later other kinds of cells too (neurons are another name for nerve cells). All this is happening in a tightly controlled manner, such that cells of different types appear only at certain times and places – as if each one has its own address. So while you can have all the right brain cell types present in a fully-grown brain – and in the right proportions – if they are in the wrong place, then the brain won't work as it should.

Ever get lost driving somewhere even with GPS?

We have.

Imagine – you are traveling somewhere but instead of moving in two-dimensions (like on a road or across the sea in a boat), you have to navigate in three dimensions to arrive at your proper destination. In fact, it's more like four dimensions, since you also have to arrive just on time or bad things will happen. That's what it is like in the developing brain.

The different types of cells of the developing brain have different hangouts, destinations, and homes. For example, brain stem cells like to hang out in certain places, such as near fluid-filled spaces called the "ventricles" and also near blood vessels. We call these places "stem cell niches." If a stem cell wanders off, they might either self-destruct through apoptosis (see Chapter 2) or develop into a more mature cell type, such as a neuron. A certain number of stem cells leave their niche on purpose to generate the mature cell types of the brain.

As the brain grows, it needs to make three key cell types, called neurons, glia, and oligodendrocytes. Oligodendrocytes (yes, they are a mouthful, we know) are cells that wrap around nerves and form a protective coating called myelin, which is needed for many types of nerves to properly transmit nerve signals. These signals are altered when oligodendrocytes are lost in certain diseases. And the complexity continues because within each of these three major cell types there are lots of subtypes. For instance, there are thought to be hundreds of different types of neurons.

As the stem cells in the brain divide and populate it as it grows, the millions and then billions of cells they produce go on to form trillions of interconnections with each other called synapses. While these synapses are just as crucial for normal brain function as having the right numbers of different cell types in the right place, these interconnections aren't fully understood. Even as our knowledge of the brain rockets forward, big gaps remain in our understanding – as a result, the cause of many common neurological conditions, such as autism spectrum disorder, remain enigmatic.

Growing a mini-brain or "brain organoid" in the lab

Recently, scientists have devised a new way of making miniature versions of just about any human organ, including large portions of the brain, in the lab by using stem cells.

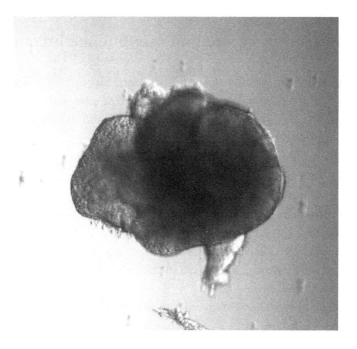

Figure 4.6. Early human brain organoids made from stem cells in the Knoepfler lab by medical student fellow Jacob Loeffler.

These mini organs are called organoids and are beautiful structures (see a human brain organoid in Figure 4.6 that was grown in the Knoepfler lab). However, we don't think that we can use organoids to generate a complete, functioning human or dragon brain any time soon, and we are relatively impatient to press on with making our dragon.

Brain organoids or "mini-brains" can be made from special stem cells in the lab by growing these cells on a plastic plate (called a culture dish) and exposing them to factors that are known to stimulate brain-like development. In fact, my lab (Paul speaking, here) grows human brain organoids as part of our research into human microcephaly and brain tumors. We mostly make them from a cool type of stem cell called "induced pluripotent stem cells" (for more on these cells, see Chapter 6). But these mini-brains are incomplete structures and are nowhere near as complex as the real thing.

Are brain architecture and intelligence linked?

So, intelligence. We have some questions about it. To begin with, how do different brain characteristics contribute to one's intelligence? And how does brain development – during gestation and after birth – relate to intellect?

When it comes to brain size, it's true to say that larger brains are generally associated with higher intelligence (as discussed earlier in this chapter). However, this isn't always the case. For instance, in rare cases some individuals with microcephaly or those who have lost a substantial amount of their brains due to injury or disease, surprisingly still have entirely normal intelligence. Conversely, an abnormally large brain can sometimes be associated with cognitive problems or decreased intelligence.

Another potential measure of brain power is the total number of neurons a brain contains. A simple way of thinking about this is to consider neurons as microchips in a computer. More chips often mean more computing power, and more neurons often equal more brain power. Still, this analogy doesn't always hold up. For example, the cerebellum (which again sits at the back of the brain and controls our coordination) contains the most neurons of any brain region.

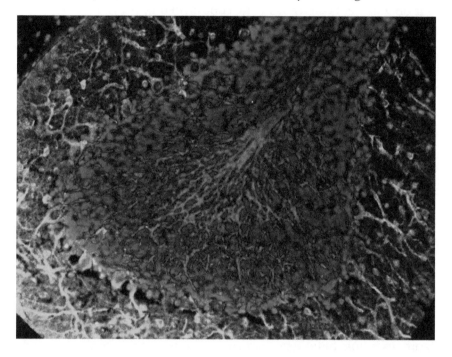

Figure 4.7. A photograph taken on a microscope of one small part (called a folium defined as "a thin, leaf-like structure") of a mouse cerebellum. The green color marks a special type of neuron found in the cerebellum called a Purkinje neuron. Red marks brain cells called oligodendrocytes. The blue color marks the DNA of each cell. Image credit Knoepfler Lab.

However, there are some extremely rare cases where humans have functioned sufficiently well to survive without a cerebellum. This means that total neuron numbers might not always provide a link to intelligence. In part, it depends on where the neurons are located and what specific type of neuron they are. You can read about the case of a Chinese woman with no cerebellum, a condition called cerebellar agenesis, in the study referenced here [2]. While she had some issues, such as movement problems, she reportedly functioned surprisingly well.

Note that only nine people in history have been known to survive without a cerebellum.[x] And we can't imagine our dragon getting along without one, even if it isn't technically always essential for life, because our dragon will need to be highly coordinated to fly.

You can better appreciate the complexity of the cerebellum in a microscopy image taken by my (Paul's) lab, from our studies of cerebellar development in mice (Figure 4.7) [3]. The bright colors come from special fluorescent molecules that bind to proteins found in the different cell types of the mouse cerebellum. Each color tells us what kind of cell we are looking at in a given image. The red colored cells are oligodendrocytes, which, as we explain above, wrap up neurons in a protective layer. The green stained cells are a type of neuron (called the Purkinje neuron) that are abundant in the cerebellum and help regulate body movement. The blue stain is for DNA and it highlights the nuclei (the structure that DNA is packaged into) of the different cells.

Another possible way to measure and predict intelligence is by neuron density. Just as the brain to body ratio (how big is the brain relative to the size of the body) is important for intelligence, so it seems is neuron density, meaning how many neurons per given area of brain. For example, although an animal might have what seems to be a small brain, its whole brain (or a certain part of it, such as its cortex) might be packed tightly with neurons. And a cortex packed with neurons is thought to be a marker of intelligence – as such, an animal with a small brain might be smarter than we imagine and more intelligent than another animal with a larger brain but many fewer neurons per given area such as a cubic inch.

And recent research has backed this up. Clever birds, such as parrots and crows, can have as many neurons (and sometimes more) packed into their forebrains as do some monkeys with larger brains. Indeed, these birds' small brains might pack more "cognitive power" than do the brains of some mammals [4].

---

[x] https://www.newscientist.com/article/mg22329861-900-woman-of-24-found-to-have-no-cerebellum-in-her-brain/

The organization of the brain is also likely to be key for intelligence. For example, the number of those connections among neurons, again called synapses, in certain special brain regions might disproportionately affect intelligence as well. And looking beyond neurons, the other cells of the brain influence intelligence too. For instance, researchers have found that when chimeric mice are created to have a mix of both mouse and human brain cells, they become noticeably smarter when just supporting human brain cells (called glia), and not neurons, are present in the mouse brain [5].

What all this means is that engineering our dragon's brain is going to be a big challenge – and that it'll be hard to get it "just right."

Too dumb

Another way our dragon-brain building could go badly awry is if our dragon ends up being just not smart enough, possibly because its brain is too small, not properly structured, has too few neurons, or has other issues. Assuming our dragon is just barely smart enough to learn to fly (and land) – and to not constantly cook itself with its own fire breathing – it might still not be smart enough to function quite right as a dragon. We must be able to teach it more than flying, as there are many other basic skills it will need, and we cannot teach our dragon all these important things if it is just too simple to understand us.

What other things might our dragon need to learn? Well, geography to begin with, so it doesn't get itself (or us) lost whilst out flying. It also needs to be smart enough to learn restraint on the fire front, and to not go out pillaging, at least not on its own – although, admittedly part of getting this just right could relate more to personality than to intelligence.

Unfortunately for us, many of the creatures that most resemble dragons in various ways – such as dinosaurs, birds and lizards – had or have small or simplistic brains (again with some exceptions for certain bright birds). The first ever fossil of part of a dinosaur brain was discovered just a few years ago. It belonged to a colossus called

Iguanodon.[xi] We don't know how they found this fossilized part of a dinosaur brain because it is so small.

Considering their huge overall size, iguanodons had a relatively minuscule brain. We think that this kind of brain-to-body-mass ratio of some dinosaurs probably made them so remarkably dumb. Again, the brain does tons of stuff to run the body and, as a result, much of the brain's effort (and mass) can end up being devoted to daily housekeeping-type things, which means for iguanodons there wouldn't be much brain left for intellect.

For all these reasons, when building our dragon, we need to err on the side of a slightly larger or more complex brain to avoid it being short of the smarts — if it was, it could foul up our plans in many ways.

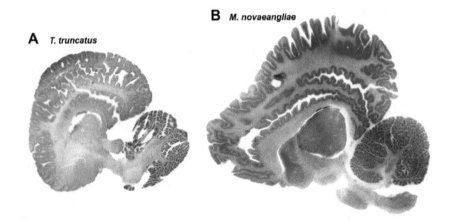

Figure 4.8. Sections of brains from a bottlenose dolphin (A) and humpback whale (B) stained with a special dye. Note the large size and complex folding, which suggests an unusually high degree of intelligence. In both cases the front of the brain points to the left and the cerebellum is on the right. Source, Marino, *et al. PLoS Biology*, 2007, an open source publication with a Creative Commons license policy [5].

[xi] http://sp.lyellcollection.org/content/448/1/383

Too smart

As we alluded to earlier, there are some animals that have bigger brains than humans, including elephants, whales, and dolphins.[xii] They also have other brain attributes pointing to high intelligence. But it's hard to know exactly how intelligent these creatures are given two factors: their large body masses and the fact that the cetaceans use a large portion of their brains for echolocation. Even so, some studies suggest that cetaceans (that is whales and dolphins) are on the extremely smart side of the animal kingdom's intelligence spectrum. For all we know, they could be more intelligent than humans in some ways [6]. The fact that we cannot easily communicate with these big-brained animals is another obstacle to judging their intelligence.

As mentioned earlier in this chapter, one part of the brain strongly linked to intelligence is the cortex. In particular, the surface area of the cortex is strongly related to intelligence, and this can be increased greatly by folding. When you look at the brains of intelligent creatures, like humans, whales, and dolphins, you will see that they have lots of folds in their cortex (Figure 4.8). That folding equals greater brain power because there's a lot more brain packed into those folds. By contrast animals generally thought to have less brain power like mice have smooth brain cortices (plural of "cortex"). They get along fine that way but aren't exactly geniuses.

In the previous section we were worried about making a dragon who is low on brain power. At the other end of the spectrum, if we intentionally or accidentally make a dragon that ends up being overly intelligent, we could be in for real trouble. The most likely disastrous outcome of creating a dragon genius is that it will find us to be too much trouble and simply kill us. Or, our uber-smart dragon might just fly away and never return, seeking its independence from humans.

Even if it sticks around and doesn't kill us, it might just decide it doesn't feel like acting the way we think a dragon should. Imagine a brainy, vegan dragon using its remarkable fire-breathing capabilities only for roasting vegetables, while spending its days reading books,

---

[xii] https://blogs.scientificamerican.com/news-blog/are-whales-smarter-than-we-are/

meditating, advocating for world harmony, making advances in theoretical physics, or watching "educational" YouTube videos. Bah!

Even if we hit the intelligence "sweet spot," our dragon could end up with any number of unwanted, weird, or pathological personalities. It's not hard to imagine an awesome dragon becoming rather narcissistic. Perhaps it'll spend its days taking selfies to post on social media, or staring into the mirror admiring itself.

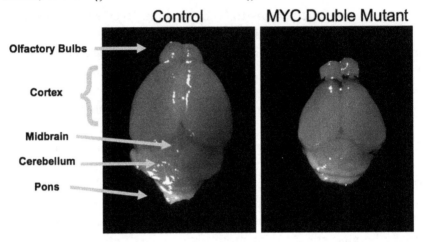

Figure 4.9. A normal (control) mouse brain (left), and a brain from a mouse of the same age that lacks two Myc family genes, *c-Myc* and *N-Myc* (right). Both images are taken using the same magnification and provide a top view of each brain. The different parts of the normal brain are labeled with arrows. Note, that the mouse lacking the *c-Myc* and *N-Myc* genes (the MYC double mutant) has a much smaller brain than a normal mouse, although all the various key parts of the brain are still present to some degree. Image credit: Knoepfler lab.

It's also not such a stretch to imagine our dragon becoming an extremely enthusiastic killer, perhaps even of humans. After all, in many (if not most) artistic or mythological works, dragons are instinctual killers. Our dragon could even end up being outright evil. What would we do then? Some of these inclinations and behaviors could be instinctual and difficult to change. The biological basis of instinct isn't well understood, but is thought to be attributable to certain genes.

J.R.R. Tolkien recommended to "never laugh at live dragons" and we don't plan to. But we believe our new dragon would need a very good sense of humor too, especially once it realizes that we two puny humans are its creators. Unfortunately, there isn't much science on how a sense of humor develops in different kinds of brains.

Extra Smart on purpose

Hopefully, you now get the picture. We want our dragon to be smart and trainable – so for one thing, it doesn't kill, incinerate or eat us – but what if we went further and gave our dragon human-level smarts? This could be difficult given that we scientists don't really know the genes involved in intelligence (and there are likely to be loads of them). Still, we do know of some genes that influence brain growth and the folding of the cortex. If we activated these genes, at the right level and in the right place, we might just be able to give our dragon substantially more brain power.

I (Paul) have studied brain growth for much of my career and worked to understand the functions of some of these genes. One such family of genes is called the *Myc* family. They orchestrate normal brain growth. If these genes are removed from the brain stem cells of mice, the mice end with smaller brains and heads than they should (see Figure 4.9) [7].

Furthermore, I discovered years ago that moderately increasing the levels of one particular *MYC* family gene, called *MYCN*, in mice led to their brains becoming more-folded, making their brains more similar to those of humans (a fascinating finding, although one I didn't publish because the mice never made it through development, most likely due to this condition being incompatible with life). These kinds of brain changes – to create a large or more powerful brain – could, in theory, be induced in the brain of our future dragon as well.

However, there would be some major risks to attempting this.

For example, our dragon could end up developing brain tumors instead of just a larger brain (and note that *MYC* genes are so-called "oncogenes" or cancer-causing genes because they are not just linked to normal brain growth but also to a variety of brain tumors and other cancers). Or our dragon's brain could end up being way too big, resulting in a condition called macrocephaly, which is paradoxically sometimes associated with altered or impaired intelligence.

Even if we tried to make a super-smart dragon, and we got it just right managing to make our dragon a genius, it might just fly off, as we've said, and do whatever the heck it pleased. But it could also pull a role reversal where we become its pet!

However, we might just mess it up and give our dragon a small brain instead. When monkeying with genes and biology, unexpected consequences become, paradoxically, more expected. Trying to influence our dragon's brain toward a certain intelligence will most likely come down to a quite dangerous game of trial and error. Biology is extremely complicated.

In myths and legends, dragons were often portrayed as extremely intelligent, perhaps equal to humans in some cases. Although often seen as evil or dangerous, dragons were also often portrayed as wise and possessing very long memories. For these reasons, if our dragon ends up being on the smart side it will not only be useful, but also generally fit with people's notions about dragons more generally.

Walk the walk, and talk the talk

We need to communicate clearly, back and forth, with our dragon, which requires a fairly sophisticated brain function and is no easy task for most creatures to do. For instance, I'm not sure that anyone has managed to train their pet reptiles to communicate. Iguanas or Komodo dragons might communicate with each other in a basic way – by hissing or with tail slaps. But we need our dragon to be on a much higher plane. It could communicate with us in English, which would be ideal (although giving it vocal cords that can both speak and roar monstrously could be difficult to do.) Alternatively, we and the dragon could converse in some made-up language (so that we can talk privately – sounds cool but a lot of work) or in some other way, maybe using sign language.

Not only does this kind of two-way communication require high-level brain function, but also a certain kind of tongue structure. So it is encouraging that some birds can both sing complex songs and actually talk in English and just about any other language (and we don't just mean the stereotypical parrot saying, "Polly wants a cracker!" or "Pretty Polly!"). Incidentally, you can read one historical theory as to why English speakers have referred to parrots as "Pol" or "Polly" for hundreds of years in this piece by Mental Floss.[xiii] The piece also has a funny anecdote about a presidential parrot named Pol who had a bad habit of swearing.

We discussed before how birds might provide us with a good creature to start building our dragon from, given their flight skills and genetic ties to dinosaurs. Maybe then a "bird brain" might not be so far off from the kind of dinosaur brain that we need, and we might even pair it with a bird-like tongue to enable speech.

If we can have a productive discussion with our dragon, it could help us avoid many potential problems. For example, imagine our dragon named "Polly" saying to us, "Polly wants to burn down the village and eat the people like so many crackers!" and then we reply, "You know, Polly, not today. If you do that kind of thing, you're not going to be long for this world as people are likely to get upset. They'll probably want to kill you. Plus it's not a particularly nice thing to do to attack the village."

Our dragon won't be able to talk on the day of its birth or creation, but rather it's going to take a lot of work by us, and maybe even others, to raise our dragon (more on that below).

Teach your dragon well

One thing that we humans have worked out over the years is that biology is not destiny. Just because you are born with a genetic program to develop a certain kind of brain (whether not so smart or a potential genius), it doesn't mean it will turn out that way.

Sure, severe conditions – like microcephaly or a traumatic brain injury – can have permanent and life-changing effects on a person's

---

[xiii] http://mentalfloss.com/article/55350/why-do-we-call-parrots-polly

brain. But for every person born, the environment they develop and live in from day one will have a profound impact on their intelligence as well as personality too.

For instance, if the famous Nobel Laureate Marie Curie had had different parents or teachers, a different diet, different illnesses, or different education, it's very possible that in some of those hypothetical scenarios she would not have had the same fierce intellect and impact on our history. (Thank goodness she turned out the way she did!) And it's likely going to be the same for our dragon.

This is because – from a dragon-making perspective – although we can hypothetically do everything right in terms of creating an optimal brain for our dragon, this alone will not be enough. From the time our dragon is born, we have to be devoted to teaching it and to providing it with the "right" environment to ensure its healthy development.

Our dragon, especially at first, will basically be like our kid, providing us with all the same opportunities that human parents have to mess up their kids. We two authors are parent and child, so we know first-hand how important it is that children have a certain environment when growing up. Not only will the manner in which we interact with the baby dragon impact its brain, but it'll also have a big influence on its overall health and well-being.

Some specific examples come to mind. For instance, assuming our dragon has the physical and mental abilities to talk, we'll need to work at teaching it to talk and provide it with an environment where it can soak up the language (or languages) we want it to speak. We can imagine spending countless hours chatting with our baby dragon, reading books to it, and maybe even homeschooling it. After all, could we actually send our dragon to a regular school? We'll need to teach it many other basic things as well including how we want it to behave.

As mentioned in Chapters 2 and 3, we'll somehow also need to teach our dragon to fly and to breathe fire without killing itself.

Problems bringing up baby Dragon

While we work to teach our young dragon all it needs to know, we expect all kinds of craziness to be taking place. A young dragon may have limited impulse control. Even just raising a puppy (like our dog Mica, see Figure 5.4 in Chapter 5) can be a wild experience.

As we were raising Mica, her puppyhood happened to coincide with us writing this book. Along the way, we'd have some crazy (and sometimes funny) experiences with Mica that made us think about what raising a young dragon would be like and the challenges that might arise.

While Mica liked to chew on small toy bones, we imagine our baby dragon needing to chew too. Maybe a growing dragon would want to chew on giant bones, perhaps from some huge animal like a muskox. Sometimes when Mica got tired of chewing on bones or her toys, she'd chew on the furniture. Might our young dragon literally chew on our house, causing gaping holes in the walls? Or the dragon might chew down full-grown trees in the backyard? Accidentally eat our entire television set?

Mica has also been very interested in human food as we cook or eat. With our smart young dragon, might it open the fridge and help itself to the entire fridge full of food? In Chapter 3 we discussed the diet of our dragon and how that might impact its powers including fire-breathing.

All that food has to go somewhere too once its digested. Sometimes on rare occasions Mica would have accidents in the house. With a young, ravenous dragon eating tons of food, such accidents could be like disasters. The dragon could leave liters of pee or giant poops weighing 10 or more pounds (ca. 5 kg) in the house. Would we need hazmat suits and giant biohazard waste bags to clean it up?

We have also found that Mica is much more well-behaved if we give her loads of exercise, mainly long walks around the neighborhood. The same kind of thing will likely be true with a dragon. If it has a huge amount of unspent energy, it's more likely to get into mischief and dragon mischief could be awful.

The more exercise the young dragon gets, the more it'll take naps and behave better when awake, but can we safely take our dragon for neighborhood walks? We can envision all kinds of problems. For example, even the "alpha dogs" of the neighborhood who are big and tough, might get afraid and upset by a dragon walking by. The owners aren't likely to react well either. Our baby dragon might see smaller dogs as potential prey! Once it learns to fly, that may be a more practical way for it to get exercise.

Unfortunately, there have been times when Mica has dug big holes in our backyard and even in our garden, uprooting some of our vegetables. If our young dragon got digging into its head, it could end up making crater-sized holes in our yard or accidentally unearth (and break!) city water pipes.

A couple of times Mica got sick, so we called the veterinarian. Would our vet or any animal doctor take a dragon as a patient? Once Mica even needed eye drops, which she hated and sort of fought us as we were putting them in her eyes. Try putting drops into the eyes of a cranky young dragon.

Another time Mica somehow ate a few grapes and it turns out that grapes are toxic for dogs, so we called the vet. Fortunately, she was fine. Then we started imagining all the weird things a young dragon might eat and found an article about all the strange things that have actually been found inside of sharks.[xiv] With a young dragon, imagine calling the vet to ask not about grapes but instead, "Our dragon ate an entire chicken coop. Is she going to be okay?"

All of these potential mishaps got us thinking about insurance. Would any insurance company be willing to insure our dragon? What about our dragon-making laboratory? We hesitate to even bring it up with our current insurer!

[xiv] https://www.sharksider.com/14-weirdest-things-sharks-eaten/

Cheating with a Cyborg Dragon

As an alternative to biologically enhancing our dragon to have solid intelligence, we could go the cyborg route and boost our dragon's brain power with a computer implant into its brain. In theory, we might even be able to "tune" our dragon's brain power remotely *via* such an implanted chip. But would this make our dragon more of a remote-controlled creature? We don't know, but it's an interesting if somewhat disturbing idea. And, yes, it would be a form of cheating. Admittedly, brain chip implant technology isn't yet established.

We explore these and other ideas for dragon features further in the next chapter.

What could go wrong – and yes, you've guessed – lead to our deaths?

We've already covered some potential problems of making a dragon that's too smart – it fries us, or drops us from the sky, or crushes us for being too stupid or annoying. And then there's the dragon that's too dumb – it might not even realize it has turned us into ash with its fire-breathing. This dodo dragon could drop us from the sky by accident or sit on us.

Trying to get the dragon brain just right

And as we've discussed earlier, even if we get our dragon's intelligence just right, it could be a pathological beast fixated on the idea of murdering us or on the other end of the spectrum a peace-loving dragon who literally wouldn't hurt a fly. Getting its brain "right" especially by way of its personality is going to be nearly impossible, but we've certainly got to try our best.

References

1. Hazlett, H.C., *et al.*, Early brain overgrowth in autism associated with an increase in cortical surface area before age 2 years. *Arch Gen Psychiatry* 2011. **68**(5): 467–476.
2. Yu, F., *et al.*, A new case of complete primary cerebellar agenesis: clinical and imaging findings in a living patient. *Brain* 2015. **138**(Pt 6): e353.
3. Wey, A. and P.S. Knoepfler, c-myc and N-myc promote active stem cell metabolism and cycling as architects of the developing brain. *Oncotarget* 2010. **1**(2): 120–130.
4. Olkowicz, S., *et al.*, Birds have primate-like numbers of neurons in the forebrain. *Proc Natl Acad Sci USA* 2016. **113**(26): 7255–7260.
5. Han, X., *et al.*, Forebrain engraftment by human glial progenitor cells enhances synaptic plasticity and learning in adult mice. *Cell Stem Cell* 2013. **12**(3): 342–353.
6. Marino, L., *et al.*, Cetaceans have complex brains for complex cognition. *PLoS Biol* 2007. **5**(5): e139.
7. Knoepfler, P.S., P.F. Cheng, and R.N. Eisenman, N-myc is essential during neurogenesis for the rapid expansion of progenitor cell populations and the inhibition of neuronal differentiation. *Genes Dev* 2002. **16**(20): 2699–2712.

Chapter 5

From head to tail, other dragon features and power-ups

A whole menu of options

Up to this point, we've focused mostly on our dragon's wings (so that it can fly), its unique physiology (so that it can robustly breathe fire), and its brain (which will run the whole "ship"). But there is a lot more to a dragon than just these attributes. From the dragon's head to its tail, there are many other features that need some serious thought. Should we give our dragon horns? More exotic features, like electrical organs? Or should it have fins to swim with and maybe gills too? And, if we do opt for such unusual features, what should they be like?

There's a lot to decide!

Even for other 'more standard' features, important decisions remain to be made.

Two legs or four?

One head or more?

A horn or multiple horns? No horn at all?

Of course, eyelids, but maybe also extra eyelids to protect our dragon's vision from fire-breathing or other dangers?

The physiological features needed for a voice?

These and many other potential features could have a big impact on our dragon's overall appearance and its ability to function. They could also indirectly influence our dragon's primary functions, including flight and that wonderful fire-breathing we want it to have.

We also need to consider whether our decisions here might impact its other attributes, like its fertility, as ideally, we want our dragon to reproduce to make more dragons. Our choices are likely to affect its overall health too, as well as its lifespan. We don't want to create a sickly dragon (even if it looks amazing) that only lives for a couple of years. And we certainly don't want our dragon to be miserable or to suffer unnecessarily.

Figure 5.1. A giant sculpture of a three-headed dragon in Russia. Image from Shutterstock.

In this chapter, we go through all the various features we've considered for our dragon and their implications. We also discuss various body parts and their associated functions, some of which exist in real animals.

We then end this chapter with a fun section discussing a range of possible dragon enhancements – what we call "power-ups."

Since these enhancements could give our dragon boosted (or even super) powers, we could create an even more amazing dragon. But a bunch of higher risks come along with enhancements including the fact that we might screw something up or get killed along the way. Still, these enhancements are going to be hard to resist so we'll at least consider adding some of them.

A head (or heads) up?

Starting up top, how many heads do we want on our dragon? Frankly, life would be simplest if our dragon just had a single head. Sticking with one head, with its accompanying one brain, would be the safest way to go.

Because there are some potentially big problems that come with having more than one head, a dragon endowed with multiple brains (inside its multiple heads) could end up with difficulties. For instance, it might have multiple personalities, which could get into conflict with one another at times.

Let's go left! No, let's go right!

Let's go burn down Toledo! No, let's go after LA or Tokyo first.

I like Julie and Paul. No, they are annoying, let's barbecue them and eat them now.

Hopefully, you get the picture.

Flying could also be much harder with say three heads instead of one. There's potentially increased wind resistance (due to the multiple heads) and navigational challenges as well. Which head will be in charge of navigation during flight? Three heads would also require a lot of energy and so would change our dragon's metabolism, requiring more and different foods to be included in its diet. Those three heads and their associated necks would also add to the dragon's weight, making it hard in that sense for our dragon to fly as well.

For these reasons, one head would be the prudent choice. However, we can't be expected to always only go the prudent route, right? Also, over the centuries, dragons have been depicted with more than one head, with often cool effects. It's therefore tempting to give our dragon an extra head or two.

A dragon with multiple heads could also have some distinct advantages – clearly, there are reasons why multi-headed dragons have been depicted in art and mythology. It's likely, for example, that a dragon with two heads would be far harder to kill than a dragon with only one. You chop one head off (or the dragon seriously damages it with a clumsy landing or by not ducking when entering its mountain lair) and yet you still have a living dragon, thanks to its remaining head and brain, which allow it to continue to function like a dragon.

And there are other possible advantages as well. Each of our dragon's heads could have a different and useful function. One head could breathe fire, for instance, while the second head could breathe storms, or shoot out electrical beams, or do other amazing things, like spit poison (more on this later in the Chapter). Or one particular head might be more resilient to "thwacks," while the other head is smarter, wiser or has better vision or hearing. These possibilities make the idea of creating a multi-headed dragon very tempting.

But how would we make a dragon with multiple heads in the lab? Biologically, it would be much easier for us to make a dragon with only one head. This is because all vertebrate animals are programmed to make just one head during their development. And evolution shaped this one-headed program for a reason – likely because of the serious kinds of problems that could occur that we've already discussed.

Still, for the moment at least, let's hypothetically say we give in to temptation and go for building a dragon with more than one head. Does such a two-headed version of any vertebrate creature already exist in nature? The simple answer is yes – there are rare but real creatures, like snakes and reptiles (and even a few humans) that end up with two heads, a condition called "polycephaly," which literally means multiple heads.

How do some creatures end up with polycephaly?

Surprisingly, it results from a unique form of a more well-known condition, called conjoined twinning, where twins become fused together while developing in the womb. The end result is a single body that has various parts from two different siblings. In some conjoined twins, certain organs are present twice (one from each), while other organs can be present only once and are shared.

Humans that result from conjoined twinning used to be called "Siamese twins." This name came from a famous circus performer (or more accurately performers) – two conjoined siblings (Chang and Eng Bunker), who became fused together in the womb during their development. They were born in a country called Siam – which is now known as Thailand– hence the name, Siamese twins. Chang and Eng were relatively healthy compared to other conjoined twins and lived a comparatively normal lifespan.

Conjoined twinning is extremely rare most likely because many conjoined twins don't survive during pregnancy or die within hours of birth. The conjoined twin embryos of humans and of other creatures often don't survive because their development is so profoundly disrupted as a result of one body becoming fused to and mixed with another.

Figure 5.2. Image of Janus, the two-headed turtle, who is a living attraction at the Natural History Museum of Geneva. He was born there in 1997. Image Credit Philippe Wagneur, Natural History Museum of Geneva (MHNG). Open source image.

Intentionally generating multiple heads in a living, generally healthy, dragon would be very difficult. In addition, even if we achieved two heads on one conjoined twin dragon, we would need the conjoined twin dragon to be symmetrical, in the sense of the left and right sides being nearly identical and with even placement of the heads on each side. However, conjoined twins often are not symmetrical on their left and right sides, which causes mobility and other problems [1]. If we accidentally make an asymmetrical two-headed dragon (e.g. the left head is up top where it should be, but the right head is sticking out the side of the body closer to where the rib cage is, or the two heads are radically different in size), it would likely not be able to function normally.

Although the exact causes of conjoined twinning aren't well understood, there are two possible ways that it could happen. Conjoined twins might start their development as two different fertilized eggs, which initially grow as two separate embryos in one womb. These embryos then fuse together early on during their development and from then on, their developing bodies become intertwined. These would be fraternal (non-identical) conjoined twins. However, it's also possible that conjoined twins start off as a single fertilized egg, which early on forms just one embryo. This single embryo then starts to split apart, as happens when identical twins form, but this splitting process doesn't finish successfully. This incomplete splitting creates two joined, but only partly separated, identical twin embryos, which then develop further in a connected form.

Polycephaly is a specific, even rarer type of conjoined twinning, in which the conjoined twins share one body, but each retains their own, separate head. There are two main forms of polycephaly – in one form, the heads are completely separate, in the other, the heads are somewhat fused together. In Figure 5.2, you can see a photo of Janus the turtle, who had two completely separate heads and who was named after the Roman god Janus, who had two faces. In another, more unusual form of human conjoined twinning, the resulting person can appear to have two faces on one head.

Since the causes of conjoined twinning and polycephaly are unclear, we have no idea how we might trigger and then use conjoined twinning to achieve the specific kind of polycephaly we'd need to create a two-headed dragon. Plus, we'd need our dragon to develop two complete, again symmetrically arranged heads (incidentally, regarding the earlier section on the importance of symmetry, notice how Janus the turtle's two heads look identical and are in symmetry to each other on the left and right sides of the front of the body). It seems like a long shot to achieve this kind of result in our dragon. However, since these kinds of things do happen in nature, even if very rarely, it's at least theoretically possible.

INDEFESSA GERENS REDIVIVIS BELLA COLVBRIS   ARGOLIS AD LERNÆ TVNDITVR HYDRA VADVM

Figure 5.3. A 500-year-old image of the Roman god Hercules, battling the Lernaean Hydra – a many-headed monster from Greek mythology. From an engraving by Cornelis Cort. This image is in the public domain.

It's interesting that the mythological beast, The Hydra (or more specifically, The Lernaean Hydra), which was often viewed as being a type of dragon, is shown in art as having an extreme form of polycephaly. Depending on which particular mythological account you read, the Hydra had anywhere from dozens to a thousand heads. You can see an engraving of The Hydra (by the 16th Century artist Cornelis Cort) battling with Hercules in Figure 5.3. In this engraving, it's shown having many heads and legs. You might note that in the same image, it seems that oddly enough a bunch of crabs and lobsters are fighting on the side of The Hydra against Hercules, but we weren't able to find out exactly why they were its allies.

Some mythological dragons had the ability to regrow heads that have been chopped off. The Lernaean Hydra is described as growing two new heads for every one that was chopped off in some myths and stories, but that's not real, right?

It's true though that in the real world, stem cell biology is revealing new insights into how it is that some animals can regenerate certain parts of their bodies when they are injured or lost. Stem cells are kind of like the "wild cards" of the animal world in that they can sometimes grow brand-new tissues. There are also real animals, like a small invertebrate marine creature, also called a Hydra, as well as some amphibians and lizards (a potential dragon starter creature) that can regrow (that is, regenerate) parts of their bodies, like a tail or limb *via* stem cells. Two of the most impressive regenerators are the amphibian called an axolotl, which is able to regrow its limbs, and the planarian worm that can regrow its whole head or even almost its entire body.[i]

However, regenerating an entire dragon head is going to be considerably harder to do than replacing something much simpler, like a tail, or a toe, foot, or even leg. (As an aside, did you remember from the last chapter that all of us humans have tails for a short time during development in the womb? We then later lose our tails while still developing in the uterus and are just left with our tail bones. See Figure 4.5.)

A head is a very complex structure that contains the brain. Would a newly regenerated brain possess all the memories of the brain it replaced – basic things like who it was, or how it came to lose its old head in the first place? Still, we suppose it's at least hypothetically possible that our dragon could regrow a lost head from its own stem cells, and perhaps also a new brain that could take orders and accept

---

[i] https://www.npr.org/sections/health-shots/2018/11/06/663612981/these-flatworms-can-regrow-a-body-from-a-fragment-how-do-they-do-it-and-could-we

memories from the surviving (one or more) brains in the remaining heads. It'd be a long-shot though.

The regenerative power of the real Hydra marine animal is remarkable in its own right – they can even regenerate their heads (which are a lot less complex than our own and lack a true brain) [2]. The Hydra's remarkable regenerative capabilities are being closely investigated by researchers. It's hoped that we might, one day, be able to use what we learn from Hydra to help repair and regenerate parts of our own bodies. One intriguing finding to emerge from this research is that during everyday life in the absence of injury Hydra must actively inhibit any extra heads from growing on their bodies. But when they lose their head, this inhibition is lost too and a replacement head is able to grow [3].

Overall, one complete and functioning head seems the wiser option for our dragon, even if it is the less exciting way to go. So, we'll stick with just the one head in our initial efforts at least.

Taking our dragon by the horn

A horned dragon seems like a great idea to us – having one or more horns on its head could make our dragon appear rather impressive. In practical terms, horns could also be useful if our dragon rammed into other creatures or found itself in battle. We think that three horns atop its head would have the most aesthetic appeal, with the middle horn being the largest. Or, we could give it five horns, which could taper in size from the largest middle horn. The horns on the dragon's head could also trail off into smaller-sized bumps or spikes all the way down its back and onto its tail.

Cast your mind back, for a moment, to Figure 2.1 in Chapter 2, of the restored, fossilized Quetzalcoatlus skeleton, which to us brings to mind a sitting, horned dragon. Look at that bony horn-like appendage on top of the head of the Quetzalcoatlus! Its function remains unknown, but it would have given Quetzalcoatlus a ferocious appearance to the creatures of its prehistoric world. At least, we imagine it that way. Let's not forget that Quetzalcoatlus was a real-life, arguably dragon-like species, lacking only in fire-breathing, to complete the dragon-like picture.

Our dragon could also have rows of spikes (which are just specialized types of horn) that it could use both for self-defense and for attacking opponents.

But what exactly is a horn?

Later on in Chapter 7, we answer this question in some depth, but it is basically either a long, narrow protruding bone covered with skin or an area of hardened skin-like material sticking out from the body. In that same chapter, we consider other creatures that we might make, in addition to dragons, including unicorns. Sometimes what we think is a horn is actually something else entirely. For instance, narwhals are whales that are famous for their single, long horn (think unicorn horn). But this "horn" is actually a long tusk or tooth. What about a tusked dragon? Probably not. Instead of a horn, we could perhaps just as easily give our dragon antlers, but a dragon with antlers might look ridiculous!

Dragon color: it's not black and white

It can be hard to pick a color to paint your house or bedroom, or for your car or iPhone, so just imagine having to pick a color scheme for your very own dragon. It could be a tough choice! Plus, we have to opt for a color scheme at the start of the process, even before our dragon exists by kind of estimating how it might turn out. We can't just paint the dragon after its born, can we?

If we also set to one side the idea of giving our dragon colorful tattoos or a wardrobe full of colorful clothes, its overall color will predominantly come in the form of the pigmentation of its skin. We could also add extra color from its feathers if it has them. In fact, the biological process that colors skin is the same as the process that colors feathers, which is not so surprising given that skin and feathers come from the same tissue during development, as we mentioned in Chapter 2.

So how does the process of pigmentation end up giving colors to the skin, eyes, hair, and other parts of animals, including humans? Diet can strongly influence the pigmentation of some species. For example, take a look at salmon and pink flamingos, both of which are pink. The reason? They eat a lot of shrimp, which are themselves pink. Even humans can end up a light orangish color – a condition called carotenemia – when they eat too many of certain kinds of vegetables, like carrots and squash, that are chock-full of a pigment called beta-carotene. My family claims that I (Paul) had a bit of an organish hue at one point as a little kid from a phase of eating too many carrots).

Figure 5.4. Picture of the authors' dog, Mica. She is a White German Shepherd, who likely has a mutation in a protein called melanocortin receptor 1, which gives her a white (mostly pigment-less) coat. But she still has pigment in her black nose and chocolate brown eyes. Image credit, Paul Knoepfler.

However, pigmentation is mostly produced inside an animal by specialized cells called melanocytes. These cells contain a pigment molecule called melanin – so melanocyte literally means "melanin cell," since again "cyte" means "cell." Melanocytes are mostly found in the skin where they donate most of the melanin they make to nearby skin cells, called keratinocytes. In this way, the pigment that melanocytes make can spread through the skin.

Melanocytes produce melanin in a multi-step process. One of these steps depends on an enzyme called tyrosinase (an enzyme is a protein that makes changes to other molecules). Tyrosinase helps to convert the amino acid, tyrosine, into another substance, which then leads to the production of melanin pigment. When tyrosinase doesn't work properly or isn't made, its loss creates a roadblock in the melanin-production process and, as a result, no melanin gets produced. Albinos, who are humans or animals that have no pigmentation, often have a mutation in the tyrosinase gene that prevents their cells from making this pigment [4].

Since albinos lack all melanin, the parts of their body that would normally be pigmented – such as the skin – end up a different, distinctive color, which can range from white to pink and even to red (red comes from the color of blood, which can be seen under the skin in this context because it is no longer hidden from view by skin pigmentation).

There are other factors that regulate pigmentation too, which are encoded by our genes. It is mainly the variations in these genes that lead to the many colors of humankind, and of the animal kingdom as well. In fact, we humans produce different types of melanin, which affects our pigmentation in different ways. For example, the eumelanin type makes a darker pigment while a different type called pheomelanin can lead to lighter pigmentation.

More generally, all cells have something called receptors on their surface, *via* which they communicate with their environment and with each other (through molecules that bind to these receptors, called ligands). Natural genetic variations or mutations in a receptor involved in pigmentation, called melanocortin 1 receptor (MC1R), can lead to somewhat predictable changes in how much eumelanin or pheomelanin an individual can produce.

For instance, one of us (Paul) has lighter skin, freckles, and reddish hair (what's left of it) and so likely has a variant of the MC1R gene that contributes to this coloring. Coincidentally, our pet White German Shepherd, Mica, also likely has a unique type of MC1R, in her case due to a mutation (Figure 5.4). White German Shepherds typically have this mutation.

While you could easily mistake Mica for being an albino (many do), on closer inspection you can see that she produces plenty of pigment in some areas, for instance, her eyes are a dark brown and her nose is mostly black, like the noses of other dogs. Albino dogs generally have pink noses and either unusually light blue or more rarely pink (due to visible blood vessels) eyes.

So, all in all, there is a lot that biologists already know about the genes and factors that control pigmentation. As such, it seems plausible to us that we could alter our dragon's color by using genetic technologies, such as CRISPR-based gene-editing tools, to achieve certain types of pigmentation. However, we mustn't forget about our starting creature and its own color because it will also have a big impact on our dragon's color.

We don't have a strong preference as to the color of our dragon, but we lean towards its different body parts having a range of colors, as that would be most interesting. Could we make our dragon like a chameleon that can change colors to blend into its background? We're not sure, but if we could that would be interesting and helpful for the dragon.

Figure 5.5. Image of genetically modified, bioluminescent fish produced by GloFish. Image source: www.glofish.com.

In mythology and art, dragons of different colors, such as red, black or gold dragons, have been described as having particular abilities and temperaments. But whether there is any real connection between a dragon's pigmentation and its personality will have to await the completion of our project.

In Chapter 1, we discussed dragon history and in dragon history, a lot of dragons were of a bluish-green hue as they were associated with water. European dragons were often shown as being red and black to symbolize their fiery or sinister nature.

From a logical standpoint, our dragon could be colored to match its environment, such as blue for the sea, greenish brown for forested areas, and so on. Perhaps our dragon could also have bold red and black stripes to warn other animals of its dangerous nature, like a poisonous snake or insect. Either way, our dragon's color could both help it to survive and be aesthetically pleasing.

We might also use genetic tools and genetic modification to give our dragon more unusual colors. For example, a company called GloFish has been making brightly colored fish for several years. These genetically modified, bioluminescent fish fluoresce with bright colors, even in normal light, but under special lighting they shine extra brightly. Imagine a dragon with those kinds of colors (see Figure 5.5.)

A shocking suggestion

In Chapter 3, where we explored various ideas for how to equip our dragon with fire, we mentioned electric eels and their amazing physiology. Specifically, we talked about the properties of special cells these eels have, called electrocytes, which together make up what's called an "electrical organ" that produces a surprisingly powerful electrical current. If we biologically equip our dragon with an electrical organ or two, it could use these to ignite its flammable gases to breathe fire.

Even if instead we eventually spark our dragon's flame in a different way, it could be quite useful for our dragon to have electrical organs like an electric eel's. Our dragon could use these to sense its environment, or to stun prey, or to fire a massive discharge of electricity as a weapon. That last possibility would be particularly helpful if we can't perfect fire-breathing for our dragon. Having an electrical organ could also power any number of cybernetic implants in our dragon (see more in the power-ups section later on in this Chapter).

Before your very eyes

It'll be important for our dragon's survival to have excellent vision and to be able to see colors.

But should our dragon have binocular vision? This is where the two eyes are positioned on the head in such a way that they work together. In this case, the two eyes' fields of vision substantially overlap, which yields a stronger depth of vision. Binocular vision would require the dragon's eyes to be placed on the front part of its head (as we humans and some birds of prey, like eagles, have).

Or should the dragon's eyes be on the sides of its head, as seen in some birds and most reptiles? This gives creatures the ability to see a much larger overall area and in a sense kind of see two separate areas at once, one with each eye.

Given the unique needs of a dragon as a predator, we think it should have the same eye placement and binocular vision as a predatory bird.

Birds in general have large eyes and excellent vision, often much better than that of us humans.[ii] One particular bird is thought to have the best vision of any animal on the planet. According to an Audubon article by ornithologist Tim Birkhead, the Australian wedge-tailed eagle has the best-known vision of any animal, but it may come at a cost. As Birkhead writes:

> "The Australian wedge-tailed eagle has enormous eyes,
> both in absolute terms and compared with most other

[ii] https://www.audubon.org/magazine/may-june-2013/what-makes-bird-vision-so-cool

birds, and as a result has the greatest visual acuity of any known animal. Other birds might benefit from the eagle's acute vision, but eyes are heavy, fluid-filled structures, and the larger they are the less compatible they are with flight."

With this in mind, our dragon should ideally have moderately-sized eyes that provide it with good vision, rather than huge eyes to give it the amazing vision of this unique type of eagle. We don't want extra large eyes to interfere with flying.

How do scientists even measure how well animals can see? Smithsonian Magazine has an interesting piece on this kind of science if you want to learn more.[iii]

Many birds also have UV vision, which we humans lack. UV vision (that is, the ability to see ultraviolet light) allows birds to see this type of light emitted from the sun or reflected from flowers. While having UV vision could be useful to our dragon, we can't think of a good reason why.

A more clearly practical feature for our dragon to have, which can be found in some birds and reptiles, is a unique transparent extra eyelid, called a nictitating membrane. Animals with nictitating membranes often need to protect their eyes from extreme situations, such as possible damage while catching prey, which would come in handy for our dragon both when hunting and fire-breathing. Eyes are delicate structures, and we don't want our dragon's eyes to be damaged by heat and fire. These extra eyelids could also help to protect our dragon eyes as it makes fast dives in flight, by protecting them from the dust and debris flying through the air.

Half a brain on and half off

You've likely heard the expression "to sleep with one eye open," meaning to sleep lightly and be ready to react to danger. As it turns

---

[iii] https://www.smithsonianmag.com/smart-news/humans-see-world-100-times-more-detail-mice-fruit-flies-180969240/

out, some animals, including cetaceans (whales and dolphins) and some birds, literally sleep with half their brain on, and keep one eye open during sleep.[iv] As this *National Geographic* article explains:

> "Until recently, deep sleep in humans was thought to be a global condition: a person is either asleep or awake but not both simultaneously...Yet birds and aquatic mammals such as dolphins and whales display the remarkable phenomenon... [where] one half of their brain is awake, including an open eye, and the other half shows the electrical signatures of sleep. This is most likely a protective mechanism, enabling the animal to fly or swim and monitor its environment for threats with one [brain] hemisphere while the other gets some rest."

We don't completely understand how birds and cetaceans can sleep with only one half of their brain, but if we figure it out, we'll give our dragon this ability too. Indeed, it might come built-in if we build our dragon from a bird. Some people, based on their behavior at least, seem to go about their everyday lives in the daytime with only half their brains awake even if in theory everything in their cranium is active!

Dragon GPS

We'd also like our dragon to have an internal compass, much like birds do. In this way, our dragon would be equipped to navigate long distances, particularly when we're not there to help it out with something like Google Maps. For many years, we humans have observed some animals undertake huge migrations, during which they travel thousands of miles with surprising precision. Until recently, we didn't know how this was possible.

[iv] https://www.scientificamerican.com/article/sleeping-with-half-a-brain/

So do some migrating animals have an internal GPS? This was thought to be likely, but only recently have researchers figured out how it might work. It turns out that inside some birds' eyes are structures called magnetoreceptors, which are basically magnetic sensors. However, these sensors are extra amazing. They depend on light to work like a magnet that responds to the Earth's magnetic field.[v] As a result, they are thought to kind of work as a light-activated compass. We'd like our dragon to have an inner compass like this too, and if we use a migrating bird as the basis to make our dragon, we might get one step closer to this goal.

Going swimmingly

Although we have largely assumed so far in this book that our dragon will be of the Western European type – with 2-4 legs, wings, and fire-breathing abilities – for many people of the world, a dragon is something fairly different from this, as we discussed in Chapter 1. For example, in many Asian countries, dragons have some unique characteristics.

These Asian dragons lack wings and sometimes even legs, and they don't necessarily breathe fire. In fact, many ancient dragons from Asia looked like sea serpents that were portrayed as adept swimmers. If our own dragon could swim like these dragons of lore, it would open up a whole new universe to it.

In theory, our dragon could still have wings and breathe fire, but spend part of its time swimming or underwater. To do so, it might fold its wings, as do some birds that dive underwater. Penguins are able to be powerful swimmers by essentially "flying" underwater with their stubby little wings, but this probably wouldn't work with big dragon wings.

We could give our dragon fins to help it to swim, and maybe gills as well, or the ability to breathe through its skin like certain amphibians. And how about a whale spout? We could even give it

---

[v] https://www.sciencedaily.com/releases/2018/02/180207120617.htm

webbed feet as its wings will already be "webbed" with membranes. Perhaps each wing could act like a sort of fin underwater as well. There are a lot of choices here for us to think about if we want our dragon to be as elegant and deadly in the water as it will be in the air, but being able to do both could be very tricky.

A leg or more to stand on

Our dragon also needs legs, at least two of them.

It never occurred to us before but mythological dragons around the world have quite a varied number of legs, including some serpent-like dragons with no legs at all and on the opposite end of the spectrum The Hydra with a whole bunch of legs (see Figure 5.5). This means that we have more choices to make, this time about the number of legs for our dragon. And we are specifically thinking over the dilemma of two *versus* four legs, not counting the wings as legs. Although we could go for even more than four legs, we think that'd make our dragon less graceful.

The dragons in Game of Thrones – and our old friend Smaug in the Hobbit and the Lord of the Rings trilogy – are all wyverns, which are a type of dragon that has two wings (modified arms) and only two legs. To get around on earth, these dragons go on "all fours" – they effectively use part of their wings, usually the wing "elbow joint," as the equivalent of a pair of front legs to walk or crawl. It's not super graceful, but they get around. This might also sound familiar…take a look at the depiction of Quetzalcoatlus in Figure 1.4 of Chapter 1.

By contrast, if a dragon has four legs, as well as two wings, it can get around much better on land and look more aesthetically pleasing. But realistically speaking those four legs might interfere with graceful flight. Still, four-legged, winged dragons are also fairly common in mythology and art, as well as in contemporary fantasy fiction, such as in the movie *How to Train Your Dragon*.

There are good reasons that all real, vertebrate winged creatures, such as Pteranodons, birds and bats, didn't or don't have four legs plus wings (six appendages in total). It may have a lot to do with aerodynamics and weight. Also, the main, default developmental blueprint for building a vertebrate animal features just four appendages, not six. So, making six appendages instead (four legs plus two wings in one creature) will probably be somewhat harder. However, arthropods (insects) are amazing fliers, and they have six true legs and usually two pairs of wings. We can keep our options open, but we'll plan to aim primarily for making the four-legged, winged kind of dragon.

And another thing to ruminate…

As discussed in Chapter 3 on fire-breathing, it'd be useful for our dragon to have something like a rumen or gizzard or perhaps a hybrid of the two. Not only would this help it to digest, but it could also allow our dragon to use gastroliths (those stones some animals have inside their guts) to spark up the flammable gases it produces.

Dragon skin

Our dragon's skin would be its first line of defense against attack. Most dragons are scaled, like lizards and other reptiles, and this has been surprisingly consistent throughout most cultures that depict dragons. So, scales wouldn't be something that we would want to mess up (although some feathers would be an interesting skin-related addition to consider, as mentioned in Chapter 3).

A big thing we talked about earlier on in the book is the need to keep our dragon safe. While we've primarily concerned ourselves with the internal safety of our fire-breathing beast, we have yet to cover how scales could be an effective way to protect our creation from accidental self-cremation. Our first thought was that if the dragon's skin became incredibly damaged, it could simply shed its skin and grow a healthy new batch. Normal scales are somewhat resistant to heat but might not, unfortunately, be entirely fireproof, so we could give our dragon thicker scales or many layers of scales to make its skin more fire resistant. We also think that the ability to simply shed and regrow new skin would be incredibly valuable.

Figure 5.6. "Mozambique Spitting Cobra (*Naja mossambica*) spitting its venom in defense." Image credit Stuart G. Porter *via* Shutterstock.

Thick, tough scales would also protect our dragon from potential accidents, like crashing while flying, or from attack by people who might be armed with powerful weapons.

Not just spitting into the wind

How about a spitting dragon that spews not just fire but also strong venom?

Think spitting cobras.

While most venomous snakes inject their venom *via* bites into their victims – to disable or kill them – spitting cobras can spray or "spit" their venom as well. (By the way, oddly a spitting cobra's venom isn't particularly dangerous overall – it can cause blisters if it lands on your skin or in your mouth, but if it specifically lands in your eyes, things can get serious as it's been known to cause blindness.)

The fangs of spitting cobras have a unique architecture that allows them to forcefully spray their venom towards prey up to six feet away (for example, see the remarkable distance of the cobra spit in Figure 5.6). The goal of such spitting is self-defense or to incapacitate an animal that might be a threat or that could be eaten for lunch. Spitting cobras have a remarkable aim, ironically called "blinding accuracy," in terms of hitting their prey's eyes, although they may be aiming more for the head.[vi]

Imagine a dragon that can spit venom like a cobra – it's a skill that could be useful. For instance, picture our dragon needing a break from all that fire-breathing, it could use venom spitting as an alternate weapon while its fire-breathing parts rest. Also, if an enemy is resistant to fire, it might still be susceptible to being blinded by spat venom.

Finding its voice

We think that almost no one envisions a silent or even quiet dragon and the noise it makes is instead typically a roar. However, we want, and actually need, our dragon to make other noises in the form of actual speech. At a minimum, our dragon must understand what we are saying so that we can train and interact with it, but most useful of all would be to have a dragon that could speak.

How would a dragon speak?

In fact, how do any animals, including we humans, speak?

Speech is actually surprisingly complex at the physiological level. It involves not only the "voice box" or larynx and specific brain regions, but also other body parts, including the lungs (since to speak you have to physically move air) and a tongue. In fact, for communicating vocally, we don't just need a tongue but teeth, lips and a nose too, as well as the pharynx (the cavity behind the nose and mouth).

[vi] https://www.nationalgeographic.com/animals/2005/02/news-cobras-venom-eyes-perfect-aim/

Our dragon needs something like a larynx even just to avoid inhaling its food into its airways, but we might need to make some modifications there and elsewhere to help it to achieve speech. Humans can talk and sing mostly because of our vocal cords, which are in the larynx, plus our brain must work to understand and interpret speech as well. Instead of a larynx, birds have a similar kind of structure called a syrinx, and they too have unique brain structures that allow them to sing and even talk in some cases, as well. Other body parts are important for each animal's distinctive vocalizations.

Lizards, however, cannot talk – they do not have the physiology for it. And this strengthens the case for using birds rather than lizards as a starting creature to build our dragon.

All in all, we are relying on our dragon being able to speak, so we'll take specific steps to give it a voice. While it's not such a priority, making it sing could be fun too.

Is it a girl or a boy or both?

So, if we want to make more than one dragon (which we do), we might need to create both a boy and a girl dragon because the best way of making more dragons long term – without having to start from scratch again every time – is to make a fertile pair that can then breed. However, making even one healthy dragon is a challenge and in parallel producing one or more of each sex that are also both fertile could be an even bigger challenge.

If we could only make one dragon, would it be wiser to make a male or a female? It seems wiser to us to make our founder dragon a female for various reasons.

To explain why let's look at the story of a Komodo named Flora in a zoo in England. She had never been with a male dragon, but suddenly she laid eggs and some of these actually hatched into living dragons. People were kind of stunned.[vii] It turns out that certain reptile species in rare cases can reproduce without a male being involved. Their eggs just start developing into embryos without first being fertilized by

---

[vii] https://www.scientificamerican.com/article/strange-but-true-komodo-d/

sperm – an unusual process called parthenogenesis. Most often in reptiles this will, in fact, produce male offspring, which then in theory can mate with the mother. That sounds awkward, but that's how lizards sometimes do things.

While humans and many other species have what's called the XY chromosome sex-determination system, where females are XX and males are XY (with some rare intersex variations), birds and many reptiles use a different system with W and Z chromosomes. In this system, a ZZ individual is a male and a WZ is a female. So, for example, when a Komodo female's eggs undergo parthenogenesis, they can only produce ZZ (male) or WW (which is not viable) chromosome combinations, resulting only in surviving male offspring. (By the way, dinosaurs may have used much the same WZ chromosome sex-determination system too or something very similar.)

In contrast, in extremely rare cases in some vertebrate species, for example, sharks, parthenogenesis can produce healthy female offspring, but never males. This limitation is due to the XY chromosome sex-determination system used in so many animals.

What this all means is that possible dragon parthenogenesis could yield more complicated results in terms of the sex of offspring depending on the nature of the dragon sex chromosomes. If you are interested in learning more, *Scientific American* has a nice overview[viii] of sex-determination systems in reptiles. Surprisingly, the temperature of the animal's surroundings can sometimes dictate their sex.

Would there be any reason to make our first dragon a male, or actually to make any male dragons?

Definitely.

Normal sexual reproduction (if one could call the mating of dragons "normal") is far more likely to produce live births and healthy new dragons as they grow. And while there are other ways to reproduce, including IVF (which is short for *in vitro* fertilization, because an egg is fertilized outside of the womb first and *"in vitro"*

---

means outside the body), they are not without risks and can impact an individual's health.

The more extreme reproductive possibilities, like parthenogenesis and cloning, are far less likely to be completely successful on a consistent basis without technological leaps from where we are today. In addition, sexual reproduction would boost the genetic diversity of our dragons, which would help them to adapt to their new world – it might even lead to new generations of dragons evolving novel features that we didn't even anticipate.

Some species of animals are both male and female at the same time or can change their sex during their lifetime – they are called hermaphrodites. A hermaphrodite dragon could be useful, but complex to make.

So, producing new dragons by sexual reproduction would be our ideal – it would boost their genetic diversity and fuel their evolution.

Power-ups

How about we push the design of dragons further? We could try to make dragons with more extraordinary abilities and powers that we like to call "power-ups."

These power-ups could go well beyond that of normal animals and even of conventional dragon abilities.

What if you could not only make a real dragon, but also make it far more powerful than the dragons typically depicted in art and mythology?

Why stop at just making a dragon that can fly and breathe fire? If we have the technology or can invent it, we could give the dragon all kinds of potential upgrades. These upgrades could involve different sorts of cool features, including various powers featured in movies or comic books.

Let's start with our dragon's eyes.

Earlier in this chapter, we talked about what kind of eyes we might want for our dragon. We could give our dragon some kind of super eyesight *via* a power-up. While Superman-style X-ray vision probably isn't possible, our dragon could have extremely powerful vision, even better than that of an eagle and perhaps more like that of a telescope. Some animals, such as snakes, have infrared vision so that's something we could give our dragon as well. Powerful night vision is another possibility.

How do nocturnal animals, like owls, see so well in the dark? Animals with "night vision" share several specific eye features. They usually have exceptionally large eyes, with which to gather more light, and distinctive retinal features. The retina lines the back of the eye and contains cells that respond to light. These are called photoreceptor cells, and they come in two types – the "rod" type, which doesn't sense color but is most sensitive to light, and the 'cone' type, which senses colored light.

Figure 5.7. Two Siamese cats exhibiting eyeshine at night. Image credit: Карма2. Open source image.

In addition to the retina, some nocturnal creatures like owls have a special layer of cells in their eye, called the tapetum lucidum. This special cell layer reflects light back to the photoreceptors, making more use of the available light.[ix] The end result is superior night vision. This reflective role also causes an iridescent effect, sometimes called "eyeshine," which you don't see in human eyes because we don't have a tapetum lucidum (even though we can get red-eye in photos). Eyeshine is quite apparent at night in the eyes of animals that have a tapetum lucidum – it's that odd iridescent glow you see when light is shone in their eyes. We've seen eyeshine in cats at night – much like that captured in the photo in Figure 5.7.

If you think that cats with eyeshine are kind of cool and creepy – imagine encountering a dragon in the dark, and the first thing you see is eyeshine in its huge eyes. Is there any point in even running away?

More potential power-ups come to mind too. Should our dragon have horns or spikes, like those we discussed earlier in this chapter? Sure, but these could be weaponized in powerful ways – they could have metal tips, or they could inject concentrated, quick-acting venom into the dragon's prey.

What about special teeth that can cut through just about anything? Sounds good to us. Our dragon could also be engineered with the ability to replace teeth with new ones indefinitely, something that sharks and alligators can already do [6].

And how about a tail with certain powers? Reptile tails are already powerful weapons, but a dragon tail could be strong enough to knock down a house or destroy enemies.

We could also make a water dragon, as we mentioned earlier, but go further by building one that can live above or in the water, breathing through introduced gills. In a sense, this would be an amphibious dragon that is not so different than some dragons of Asian mythology.

Other versions of dragon 2.0 are possible as well.

Extra-speedy dragons would be exciting. Pterosaurs have been predicted to fly up to 60-70 mph (over 100 km/h).[x] If our dragon could fly at far greater speeds, more like that of jets or at least Iron Man, it'd

---

[ix] https://www.nationalgeographic.org/media/birds-eye-view-wbt/

[x] https://www.livescience.com/24071-pterodactyl-pteranodon-flying-dinosaurs.html

be able to get around a lot quicker. Even a modest boost in flight speed to 150 mph (ca. 240 km/h) would be fantastic.

We've also briefly discussed the idea of a dragon that could regenerate a chopped off head, but how about a super-regenerative dragon à la Wolverine from X-Men? If our upgraded dragon could heal itself of almost any damage *via* its stem cells, that ability would be quite helpful in battle or if someone tried to sneakily kill our dragon. We'll discuss the idea of stem cell-based hyper-regeneration more in the next chapter.

An armored-up dragon could also fend off intense attacks. Maybe we'd want to make one of our dragons have robust Kevlar-type armored scales?

What about giving our dragon an extremely long life or even immortality? Even within the realm of fictional worlds such as The Lord of the Rings (remember the dragon Smaug) and Game of Thrones, dragons are neither immortal nor resistant to all things thrown at them. In both fictional worlds, dragons can be killed. More generally, immortality isn't a real thing, right? Even a sea creature called the "immortal jellyfish," which some claim can live forever, probably actually can't.

Even if we don't make our dragon immortal, an exceptionally long-life might be desirable, but we'd have to think about what it would mean for us and the world if our dragon outlived us. Who would be there to look after it and try to influence its actions? See more on these possibilities in Chapter 8.

Power downs

Although we have spent a lot of energy exploring ways to make our dragon more powerful and more invincible, we also think it would be wise for us to have a way to essentially "turn off" the beast should things go terribly awry. For instance, if our dragon is about to eat the town mayor or maybe even us, what could we do about it? We could ask it to stop. We could tell it forcefully to stop like owners often try to do with their pet dogs, but sometimes that doesn't work.

In the case of a dog that steals sandwiches and poops in the living room, it's not the end of the world if it doesn't listen to you. But if it's about to attack your neighbor, then things can get very serious, very quickly. With a dragon, the stakes are going to be even higher since they are potentially far more dangerous. What if your dragon misbehaves in dangerous ways and won't stop despite the conventional steps you take to try to put an end to its bad behavior?

If such a dragon emergency were to happen, it'd be ideal – and perhaps even lifesaving – to have a way to completely take control. For this purpose, we could have some kind of "off switch." In the most extreme case, this off switch could be fatal. We definitely wouldn't want to kill our own dragon, but we'd rather do that than have it kill us or thousands of people.

It's painful to say this, even about a hypothetical scenario, but it'd be better to lose the dragon. We aren't fools. As we've discussed throughout the book, there are many ways our plans could go terribly wrong, so we need a contingency plan for when an out-of-control dragon is heading for disaster.

You could, of course, ask – rather than using some fancy method, why not simply go out and kill the out of control dragon yourselves? Well, this could be quite tough (as well as tragic) to achieve. After all, we want our dragon to be resilient and powerful. To kill a dragon in a battle is no trivial thing. We'd likely end up being the ones who are dead. If we felt we had no other choice, a clever off switch is likely the best way to go instead.

Such a fatal off switch could be partly biological. We could design a system to release a quick-acting toxin inside the dragon or electrically stop its heart. The trigger might consist of a wireless remote control-type mechanism, which could, for example, open a small capsule embedded in the dragon to release the poison or to give its heart a fatal shock.

Alternatively, we could genetically engineer an off switch by making our dragon susceptible to a rare chemical (remember Superman and kryptonite?) that in certain circumstances would be lethal to it. But this would be a clunkier approach because we might have to trigger the genetic susceptibility somehow such as by altering the dragon's diet.

For example, while our dragon was well-behaved, we could feed it food that makes it resistant to the lethal chemical. But should it become out of control, we'd switch its diet to one that makes it sensitive to the chemical. And then we'd have to wait. Because this is a slow process!

We'd need something faster. If our dragon goes out of control, we could then give the dragon the "dragon kryptonite" chemical such as by a quick injection (like an EpiPen that can quickly inject epinephrine into people having a potentially life-threatening allergic response to something like a bee sting) or by hiding the substance in the dragon's favorite food (for instance, dog owners sometimes give their pets some medicine hidden in peanut butter or sausage). But that dragon-killing chemical would have to be non-existent in nature and only available to us.

In a less extreme contingency plan, we could also engineer a reversible "off switch" for our dragon. Instead of killing it, this switch could only temporarily incapacitate our dragon. It would be more like an on-off switch that can go back and forth.

We favor this idea over a lethal "off switch" since we'd get to keep our dragon while (hopefully) still avoiding disasters. For instance, rather than having an embedded poison or explosive capsule in the dragon, we could instead have a remote-controlled capsule filled with a knock-out drug, like an anesthetic. But we'd have to figure out what a safe and effective dose would be. And what drug we could use that would make a dragon unconscious but not harm it. This would have to be the subject of an entire research project, perhaps building on what's already known about tranquilizing other large animals, like alligators.

Even if we were successful in engineering a seemingly safe and effective "off switch" into our dragon, what if we hit the switch when our dragon is cruising around in the sky? That would not be good. So, we might have to be patient in terms of choosing when we trigger it. But as we wait for the right time, the out of control dragon could do more harm if it spends hours in the air.

And if the dragon is far away from us, we might not even know what it was doing when the time came to try to knock it out. We could try to put a permanent webcam on our dragon, with a live stream that we could access to monitor what it's doing, but what if the webcam gets hacked? Or broken? Or if our dragon manages to pry it off?

Overall, we've seen these kinds of implanted devices portrayed in spy movies. If the dragon flies far away from us, then triggering the off switch by remote is problematic – not just in terms of timing, but it might not work over long distances. We'd need to use some kind of satellite signal, so we could activate the switch from almost anywhere on the planet. This is getting complicated, right?

Another biological option to consider then is a neural implant, which we could activate *via* Wi-Fi. The implant would then send an electrical signal to the brain to make our dragon unconscious – so we'd knock it out with a small burst of electricity rather than with a chemical drug. That "knockout" electrical burst could even come from the dragon's own electrical organ inside its body (see Chapter 3). Either way, we wouldn't be killing the dragon, but just knocking it out.

One of our biggest worries about an off or an on/off switch is if someone else gets hold of the controller device. They could then take control of our dragon, and they might just use that control to destroy it. They could also use the switch to make the dragon do bad things.

What could go wrong and how might we die?

It seems that for every possible dragon trait, there is a sliding scale of risk for us as its creators and caretakers (we hesitate to say owners because can you really own a dragon?) The more extreme the kind of power we give to our dragon, the more likely it might be to turn that power against us to willfully or accidentally wipe us out. This kind of consideration makes it clear that power-ups are both exciting and terrifying in different ways.

Fun with features

Overall, even if we face all kinds of dilemmas, and we'll have to weigh certain risks *versus* potential benefits, we are going to have a lot of fun outfitting our dragon with its various features. If we make many dragons, we can experiment with different attributes from head to tail. We'll also enjoy playing with the potential of dragon power-ups too.

References

1.   Levin, M., *et al.*, Laterality defects in conjoined twins. *Nature* 1996. **384**(6607): 321.
2.   Bode, H.R., Head regeneration in Hydra. *Dev Dyn* 2003. **226**(2): 225–236.
3.   Shostak, S., Inhibitory gradients of head and foot regeneration in Hydra viridis. *Dev Biol* 1972. **28**(4): 620–635.
4.   Tomita, Y., *et al.*, Human oculocutaneous albinism caused by single base insertion in the tyrosinase gene. *Biochem Biophys Res Commun* 1989. **164**(3): 990–996.
5.   Flanagan, N., *et al.*, Pleiotropic effects of the melanocortin 1 receptor (MC1R) gene on human pigmentation. *Hum Mol Genet* 2000. **9**(17): 2531–2537.
6.   Wu, P., *et al.*, Specialized stem cell niche enables repetitive renewal of alligator teeth. *Proc Natl Acad Sci USA* 2013. **110**(22): E2009–E2018.

Chapter 6

Sex, dragons, and CRISPR

Evolving a dragon at warp speed

Do all research paths to producing a real dragon have to start with or depend on sex?

Not exactly, but from what we've learned while researching this book, we believe that the different possible research paths to building a dragon all depend – one way or another – on reproductive and developmental biology experiments. And this research will mostly rely on sexual reproduction, either spontaneous or assisted by *in vitro* fertilization (IVF). There are just a few possible exceptions to this approach, such as cloning, which we discuss later in this chapter as well.

The key to success in creating an animal with dragon-like traits, and ultimately to creating dragons, rests on us being able to alter an animal's DNA – either in whole embryos or in specific cells of an embryo during reproduction. For this purpose, we'll need a good-sized population of our potential "starter" animal (such as birds or lizards, as we've discussed in the previous chapters of this book). Then, as we build intermediate animals from them that are increasingly more similar to dragons, we'll need to accumulate groups of these intermediate creatures until we have a dragon.

What this all means is that regardless of our starting animal – whether it's a Draco lizard, bird or even some wild combination of creatures made into a dragon-like chimera – we need to routinely be able to generate and modify embryos from them. For simplicity, most of these embryos at various stages of their development will most likely be generated by sexual reproduction, perhaps sometimes naturally and at other times with a boost from IVF. This will all be taking place in our planned dragon-making lab, which will need to include an animal facility. To do this all properly and to care for the animals appropriately, we'll need to employ at least one veterinarian and veterinary technicians as well, to care for the animals and carry out the IVF.

What exactly is IVF?

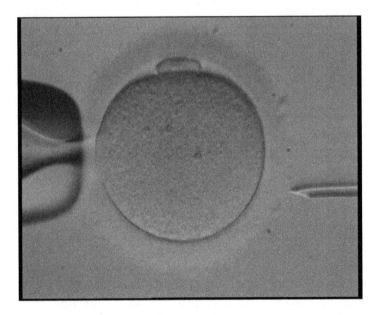

Figure 6.1. Pictured is a form of human *in vitro* fertilization (IVF) in which the egg (round object in the middle) is injected with sperm *via* a needle (right) instead of waiting for sperm to fertilize eggs spontaneously on their own. The object on the left is a tool that holds the egg in place for the injection. Image source: Wikipedia user Dovdena, creative commons license, no modification.

It's a technique that was developed 40 years ago in the UK by physician Robert Edwards to enable infertile couples to have a baby. In IVF, fertilization does not occur in the body (as it would naturally) but instead outside in a dish, which contains an egg and sperm. And just to be clear, fertilization is when a sperm binds to an egg, fuses with it and thus fertilizes it to form an embryo. It's the first step in the development of a vast number of animals, including humans. During the overall IVF process, the resulting embryos are then implanted into the mother-to-be's uterus, where one (or sometimes more) develops into a healthy baby nine months later. Sometimes instead of just mixing sperm and eggs together in a dish and hoping for the best, reproductive specialists will inject the sperm directly into the egg (Figure 6.1) to be sure the two reproductive cells join together.

The "I" and the "V" in IVF stand for "*in vitro*" which means outside of the body. This *in vitro* step for fertilization (the "F" in the acronym) doesn't just circumvent some of the key problems that infertile couples face, but it also provides a window of opportunity to make genetic changes to an embryo before it's returned to a mother's uterus, for example, by using a gene-editing technique called CRISPR. We don't believe doing this in humans is a good idea for a variety of reasons. In animals, in theory, sperm and egg, and also embryos, from different species could also be combined during IVF to try to make chimeras too.

You can read more about IVF and Robert Edwards, who won the Nobel Prize for his work, as well as its implications when used in conjunction with CRISPR in Paul's book, *GMO Sapiens* [1].

Beyond the breeding of animals, either spontaneously or by IVF, other cutting-edge technologies, such as cloning, could come into play too as we try to build our dragon. Such methods can provide more flexibility, rather than relying solely on sexual reproduction. However, although cloning should ideally generate near-perfect replicas of an animal, the process can be more unpredictable in some ways. Cloning also has not yet been achieved or perfected in many species.

In addition to cloning, some elements of our dragon-building research will likely depend on stem cells as well. There is a great deal of interest in using stem cell technology to tackle and potentially reverse infertility in humans. This kind of stem cell research aims to produce eggs and sperm from stem cells, for use in IVF. Some people have even talked about using human stem cells for human cloning, although that is a very controversial idea. In our dragon research program, stem cells could provide us with an important tool for testing different technologies and approaches, such as the editing of specific genes that we hope will create the dragon-like traits we want to make.

We're also going to need a high degree of precision when we make genetic changes to embryos, both in terms of what we do and when we do it. Successful genetic engineering can often depend on making exact changes at specific times in an animal's development. For instance, the developmental program that makes an arm might be altered precisely to make that arm look and function more like a wing, but we can't make these changes at just any old time. We will need to alter this developmental program at just the right time and place. In a similar vein, we might be able to make precise changes to an animal's gastrointestinal tract during its development to create a rumen, which we think will be important for fire-breathing, whereas this animal wouldn't normally have a rumen. However, if we make the change at the wrong time then either nothing may happen or bad things like a giant rumen could form at the expense of other tissues.

To have a chance of succeeding, many such genetic alterations will have to be made at one of the earliest stages of development. The further along that development proceeds, the less amenable embryos or fetuses are to change. However, some changes might only work if we make them much later in an animal's development – make certain changes too early and you could mess up its entire development.

Before we fully dive into the process of using genetic engineering to build a dragon, we first need to do some preliminary research on the reproductive cells and embryos of one or more creatures. Along the way, as we conduct many kinds of experiments, we'll have to deal with numerous failures before we come close to getting things right. That's just the way science works in the real world.

Fortunately, we won't have to start entirely from scratch in terms of the science of dragon building. Our work will build on decades of embryology, developmental biology, genetics, and stem cell research, even if we are taking it in a new – and some would argue entirely crazy – direction.

While quite a lot is already known about bird development, we found less published information about lizard development. We fully intend to succeed in making a dragon, but on the off chance that we fail (we think this unlikely, of course), our work could still advance the world's knowledge of these less-studied animals in some important ways, meaning that we wouldn't have entirely wasted our time.

Our efforts will also involve one of the hottest technologies around today, CRISPR-Cas9 gene-editing (often simply called "CRISPR"), which can be used to make specific changes to genes. We plan to use CRISPR to create certain genetic modifications to generate traits specifically associated with dragons. To achieve this, we'll use CRISPR in stem cells, sperm, eggs and even embryos that consist of only one cell. When these CRISPR-modified embryos are born and grow up, they can be bred to produce a living, genetically modified organism. You can read more about CRISPR later in this chapter.

For both our starter creatures and for our dragons, spontaneous sex could end up being a complicated process that slows our dragon building. What if dragons take 50 years to reach sexual maturity? Our research might end up taking more time than we researchers have to work or have left in our own lifespans. Even if we can speed things up somewhat in the laboratory, such time-consuming steps could be challenging roadblocks.

It's not always the case, but generally the larger the animal, the longer the pregnancy. Komodo dragons have an unusually long gestation period for lizards, about the same length as that of humans. So based on Komodos, a much larger dragon could have a gestation period reaching years. We don't have that time! Komodos and some other animals can also delay reproduction for long periods, and will lay their eggs only once conditions have improved.[i] Dragons may end up doing this as well, which could slow things down even more.

It's also possible that even if we successfully make fertile male and female dragons, they might not mate once they mature, or they might kill each other first as they prepare to mate. Turning to Komodos again as an example, male Komodos engage in fierce fighting before a winner is declared. These battles generally lead to much bloodshed but fortunately not to death. The victorious male then turns his attention to a female, and they can end up fighting as well [2].

---

[i] https://nationalzoo.si.edu/animals/komodo-dragon

For all the above reasons, IVF and cloning could help to speed things along.

Reptiles tend to lay many eggs. Only a few reptiles give birth to live young rather than eggs. Our hope is that when we try to breed our starter and intermediate animals, and also when we breed our dragons as well, that they'll also lay a lot of eggs. Ideally, we will need many eggs to work with because not all of them will hatch even under "normal" conditions, and this risk will increase as we start to genetically modify them in ways that might inadvertently cause their development to slow or fail. Even if we get fewer eggs if we choose birds as our starter creatures, many birds do still lay a good number of eggs.

If our dragon building plan wasn't hard enough, various lizards, including Komodos, also regularly engage in cannibalism. For adult Komodos, any younger, smaller Komodos are seen as food. Reportedly, a surprising proportion of the Komodo diet consists of other Komodos. Imagine the tragedy of making a dragon, breeding it successfully, and then watching helplessly as the dragon baby or babies are eaten by an adult dragon.

Another likely challenge in our dragon-breeding efforts is correctly distinguishing females from males. For both of our two most promising types of starter creatures, lizards and birds, it can be almost impossible for everyday people (non-experts) to distinguish their sex. One potential way around this is to sequence their genomes (or stain their chromosomes) since again male and female animals have different sex chromosomes, but for some animals, their set of chromosomes doesn't always equal their sex.

Other animals can sometimes change their sex during their lifetime, depending on environmental conditions, such as temperature, as mentioned earlier in the book. In some species, if one sex is in short supply, such as males, the females can switch to being males. These kinds of changes are more common during development, but they can occur into adulthood as well, as happens in certain frogs and fish.[ii]

What this all means is that breeding our starter creatures – and ultimately our actual dragons – is going to be far more difficult than

---

[ii] https://indianapublicmedia.org/amomentofscience/sex-nature-happen/

getting a pet mouse or gerbil to have tons of babies in a matter of weeks.

In a sense, what we would be doing with all of these efforts is guiding and speeding the evolution of dragons. We hope that a process (dragon creation) that might never happen or that might take hundreds of millions or billions of years, is being condensed down into only a few decades.

Yes, we're hoping to build dragons in a period of a few decades so that we can have them around while we are still alive.

Dragon sex ed and parenting class

If we go to the trouble of making a dragon, or ideally at least a breeding pair, would they instinctively know how to mate and produce offspring? And be good parents?

Perhaps not.

We might have to work with our dragons to get them to mate. We don't quite know how we would conduct "sex ed" classes for dragons, which could be quite awkward for everyone involved and possibly dangerous for us as well. It's awkward enough to talk to your own kids about sex, but how would anyone do this with dragons? Nevertheless, it might need to be done.

We would need our dragons to be good parents too if we're to stand any chance of generating more surviving dragons by sexual reproduction. Lizards don't "sit on" their eggs to keep them warm for hatching, and once those eggs hatch the baby lizards are generally on their own. However, we imagine our parent dragons taking care of their young to boost their chances of survival. Alternatively, if need be, we could step in as foster parents for the baby dragons.

The golden egg (and sperm)

If you think about it, it's amazing that we all grew from a single fertilized egg. But if something goes wrong in that one crucial starting cell – like a particularly bad mutation – then we are in trouble because all the trillions of cells that make up a human (or a dragon or any other animal) come from that one starting cell. Conversely, if we manage to precisely engineer a specific desired mutation in a fertilized egg – for instance, by using CRISPR – then ALL the cells of that animal will have that desired exact same genetic change. From a dragon engineering perspective that would be a good thing. By contrast, it would be nearly impossible to separately make a precise genetic change in all the hundreds of billions or trillions of cells in an adult animal, like an adult lizard.

How about changing the genetic state of all (or nearly all) the cells in just one key part of an adult animal, like a lizard arm? Well, it might be a challenge, but it would at least be relatively more doable. The best real-life example of this comes from some cutting-edge medical research – scientists have managed to make defined genetic changes to the entire immune system of some patients. These patients have a mutation (an error in their DNA) that prevents their immune system from working properly, a condition called immunodeficiency. More recently, some immunodeficient patients with this kind of mutation were potentially cured when the genetic error causing their illness was corrected in their blood stem cells.[iii]

As we discussed in earlier chapters, our efforts to make a dragon will for practical reasons likely begin with an existing creature or with a combination of creatures, rather than making a dragon from scratch. But by that we don't mean using adult animals themselves. Instead, we'll use their reproductive cells or stem cells.

On a side note, we could try to produce a dragon "from scratch" using cells and other raw materials together with technologies such as 3D-printing. However, this currently presents us with a far greater technological leap than genetically engineering a dragon from an existing starter animal. Still, we have to admit that it's fun to consider the crazy idea of 3D-printing dragons – someday it might be worth a try if 3D printing technology greatly advances. Although this might sound far-fetched, scientists have already used

---

[iii] http://newsroom.ucla.edu/releases/pioneering-stem-cell-gene-therapy-cures-infants-with-bubble-baby-disease

stem cells and other cells as the "ink" in 3D printers to try to produce actual living tissues [3, 4].

In theory, another "out there" route to building a dragon is what we call the "Frankenstein" approach, where we'd fuse together parts of other animals to build a dragon. For example, imagine a creature with the body of a large lizard but with the wings of a huge bird surgically attached. It is unclear how (or even if) our Frankenstein dragon would function as a unified whole creature, but it would definitely look scary if we pulled it off. And thinking about it, it was the 200th anniversary of the publication of Mary Shelley's *Frankenstein* as we were writing this book mostly in 2018, so we perhaps shouldn't dismiss this idea entirely.

Still, a far more realistic starting point would be to use the egg and sperm cells of our key "starting" animal as a launch pad. For instance, we could harvest eggs from Komodo dragons (very carefully, so as not to harm them or get killed). We could then use CRISPR to genetically modify the eggs in the lab to introduce a dragon-related trait and then fertilize these modified eggs in a dish (as in IVF) using Komodo sperm (also gathered rather cautiously, but we haven't figured out how just yet). These fertilized eggs would then be allowed to develop in an incubator until the eggs hatched. Inside the egg during this time would be an embryo that is hopefully somewhat dragon-like.

Also, once hatched, the resulting "test tube" Komodo dragon baby would hopefully be both healthy and perhaps also a step closer to being our flying, fire-breathing dragon. For example, our test tube Komodo might sport entirely new patagia, which could eventually help it to fly (see Chapter 2 for a reminder of what patagia are and how they aid in flight). Over several generations, and with numerous genetic tweaks, these patagia could be built into actual wings.

And what if we were to start with a Draco lizard instead of a Komodo dragon? (You might recall from Chapter 2 that Draco lizards are relatively small animals that can soar in the air but can't actually fly.) Well, we could use much the same approach. We could obtain sperm and eggs from adult Draco lizards and use CRISPR gene-editing to alter some specific genes. We might try, for example, to make it much bigger. We could also try to give it longer arms to accommodate more functional wings, or to endow it with modified (bigger) patagia. These could attach across the full length of the arms rather than only part way, as is the case in normal Dracos (see a typical Draco in Figure 2.2 in Chapter 2).

To us, some real animals already have a dragon-like appearance while still developing, such as some species of bats. In Figure 6.2, you can see images of developing black mastiff bats. As their development proceeds (from left to right), their forelimbs lengthen and their patagia grow, equipping them for eventual flight.

Biologist Dr. Dorit Hockman, who generously allowed us to use the research images in Figure 6.2, has done some cool research on wing development in bats and in other creatures as well. This work has revealed some specific molecules that regulate wing development and how they work [5, 6].

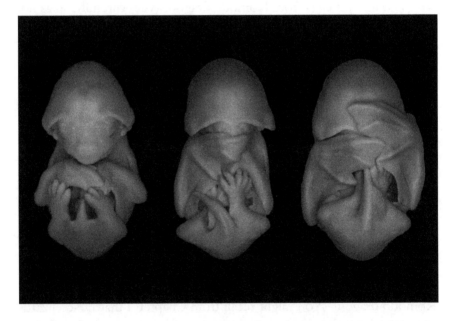

Figure 6.2. Developing embryos and fetuses of the black mastiff bat (*Molossus rufus*) at different stages of development. As fetal development progresses (from left to right), you can see how the forelimbs grow and the patagia become more apparent. To us, these developing bats look similar to how we imagine a fetal dragon might look. Copyright: Dorit Hockman. Used with permission.

One molecule that is key to bat limb development is called Sonic Hedgehog (yes, after the video game character) [7]. Sonic Hedgehog is a powerful growth factor that instructs cells how to behave, and also how much to proliferate and when to mature. This molecule is important for the development of numerous animals – from flies to humans – and works in much the same way across many animals. In biology, this consistency is described as being "highly conserved." (Paul's own earlier research on Sonic Hedgehog showed that it plays important roles in both brain development and brain cancers.)

Figure 6.3. A newborn bearded dragon hatching out of its egg, following an approximately two-month incubation period. Image from Shutterstock.

We could also start our dragon building by using a bird. To do so, we'd need to make genetic changes early in a growing chick's development while it is still inside the egg. As it happens, the basic techniques to do this already exist, as bird embryos are great for studying embryo development – they are relatively easy to do research on and to modify in the lab. If we start with birds, we might focus our early efforts on making genetic modifications that will be useful for fire-breathing. Since birds generally don't have teeth, we might also aim to give our starter bird dragon-like teeth.

In terms of our research and being able to genetically modify embryos, again it is quite useful that birds, Komodos, Dracos, and some other reptiles (see bearded dragon baby hatching out of its egg in Figure 6.3) lay eggs. This means that much of their offspring's development occurs outside of the body. It also means that genetically modified eggs can be left in their shells to develop, be incubated and then hatched in the lab. Researchers have also found that with proper training there is relatively little risk of damaging an embryonic chick when genetically or otherwise modifying it.

When I (Paul) was in third grade, my class incubated normal quail eggs. It was amazing to see them hatch in the classroom incubator, but imagine if what hatched out in your classroom was a dragon-like creature or an actual dragon!

A famous stem cell researcher, Dr. Robert Lanza, did chick research as a teenager. Apparently, the young Lanza caught the attention of Harvard Medical School researchers, when he showed up at the university one day, having successfully performed chick developmental biology experiments in his basement.[iv] (Note that it'd probably be difficult to build an entire dragon in your basement DIY lab, but maybe someone will try.)

However, it's not clear if IVF can be done successfully in birds to generate healthy eggs and then normal offspring [8, 9]. If we cannot do bird IVF, then we likely would have trouble doing CRISPR if we want to start on single-cell embryos from birds. In that case, we would either have to CRISPR the bird embryos later in development so only some cells in the animal would have the desired genetic changes or do CRISPR on avian embryonic stem cells, which would then be put into bird embryos. Overall, this could be a major obstacle to dragon production from birds, and we might have to do research ourselves on optimizing bird IVF.

---

iv http://www.robertlanza.com/who-is-robert-lanza/

The same kind of hurdle could come up while trying to genetically modified reptile reproductive cells during IVF, although researchers are working to develop genetic modification methods for reptiles [10] and for birds too.[v] It was exciting to read a report that researchers had successfully CRISPR'd lizards in early 2019.[vi]

By using the reproductive cells of our potential starter animals, we'd have a solid foundation on which to build our dragon. We'd mostly be relying on normal reproduction, combined with gene-editing and the creation of chimeras (more on this below). But because making genetic changes or creating chimeras can sometimes lead to unpredictable outcomes and to failed development, we'll most likely plan to make only a handful of genetic changes at a time. This will require making multiple generations of animals, so we're going to need to be patient.

Since this multi-generational process could take way too much time (perhaps even longer than our lifespans if we are unlucky), we are still researching ways around it. If we are going to go to all the trouble of building a dragon, we'd like to be around to see it – even if it just kills us.

It is also possible that we could generate hybrid animals by trying IVF with eggs from one species and sperm from another. Animals of different, but highly-related species sometimes successfully mate together in the wild. Unfortunately for our dragon-making efforts, these matings often produce hybrid offspring that are infertile, have health problems, and/or die before reaching sexual maturity.

Still, some cross-species matings do lead to fertile, healthy offspring. And even if we don't get fertile offspring but the offspring are otherwise healthy, we could potentially create a dragon this way and then clone it (more on this below). One of the best examples of successful interspecies mating is between different kinds of birds in the wild,[vii] so if we start with birds in our dragon-making project, we may have more flexibility on this front.

---

[v] https://www.nytimes.com/2019/02/25/science/split-sex-gynandromorph.html

[vi] https://www.sciencemag.org/news/2019/04/game-changing-gene-edit-turned-anole-lizard-albino

[vii] https://www.nytimes.com/2013/04/23/science/does-bird-mating-ever-cross-the-species-line.html

Stem cells

As mentioned earlier, stem cells are likely to be a key technology for dragon building. What exactly is a stem cell and what are the different types?

Although reproductive cells, like eggs, are technically types of stem cells, other kinds of stem cells also exist that we might, in addition, consider using in our dragon-building program.

Let's pause for a moment and talk about how we define stem cells, particularly since some readers might be surprised to hear that reproductive cells are, technically speaking, stem cells.

Stem cells have two defining properties: they can "self-renew" (that is, they can make more of themselves), and they can make other, more specialized cells through a process called differentiation. The ability of stem cells to mature (differentiate) to form specialized types of cells is also called "potency." What this means is that stem cells must be able to self-renew and have potency. Since reproductive cells, through fertilization, can give rise to a whole new creature that itself contains more reproductive cells, we define reproductive cells as being stem cells.

Most stem cells are not as flexible as reproductive cells. While they are good at making more of themselves, they mostly can only make a few highly specialized cells through differentiation. For instance, muscle stem cells can make more of themselves and make more muscle, along with a few related cell types sometimes. That's it. Blood stem cells can normally only make blood cells, lung stem cells can only make lung cells, and so forth.

In the dragon building lab, we would most likely make use of special stem cells that are called pluripotent stem cells. These unique stem cells are more powerful than the others because they can make any specialized cell type. Through the process of differentiation, pluripotent stem cells can thus make neurons, muscle, lung cells, and pretty much any other cell type in the body. They are nearly as powerful as reproductive cells.

There are two main types of pluripotent stem cells: embryonic stem cells (ESCs) and induced pluripotent stem cells (IPSCs). Both are special stem cells that can be made in the lab. ESCs are most often isolated from *in vitro*-fertilized embryos. In the case of human ESCs, for example, they are usually isolated from spare embryos generated during IVF; they can be isolated from the embryos of other species as well, like cows [11]. More recently ESCs have been produced using a cloning technique [12, 13], but the process involved is more complex.[viii]

To be clear, the cloning technique used to make ESCs is distinct from (although related to) the process used for reproductive cloning. Reproductive cloning produces an entirely new and identical copy of an organism (although in the form of a baby, not an adult), such as that often seen in movies or on TV shows, such as *Orphan Black*. Although a clone is almost identical to the creature it was made from, it might have some differences caused by the cloning process itself or by the different environment it developed in during its gestation.

If only we already had some dragon cells, we could try to use them in reproductive cloning to make more dragons. Perhaps once we've made a dragon, we can use cloning to make cloned embryos for possible future use to make even more dragons. As with human IVF embryos, we could also freeze additional dragon embryos in liquid nitrogen to ensure that our dragons don't become extinct. Such cryopreserved embryos could be thawed years later to make more dragons if there's a need.

And if we have bad luck and our dragon turns out to be infertile – for whatever reason – we could, in theory, use its skin or blood cells to make those other pluripotent stem cells called IPSCs. But before we go any further, let's first explain what IPSCs are and how they differ to ESCs. IPSCs are like ESCs in almost every way except that ESCs are made from embryos, while IPSCs can be made from any of a variety of ordinary cells found in fully developed animals that have been born (so not embryos), such as skin cells [14].

Why would dragon IPSCs be useful?

Unfortunately, while genetic engineering can sometimes yield cool new animal varieties, again the resulting offspring can also sometimes be infertile. So, if our created dragon turned out to be infertile, we

---

[viii] https://www.nature.com/protocolexchange/protocols/3117

could use it its skin cells to create IPSCs. We could then use these IPSCs to try to make more dragons by making reproductive cells from them or by using them more directly to make a dragon clone.

Even if the dragon is fertile, it might be wise to make and freeze down some dragon IPSCs as an alternative approach to propagating dragons. Having a cryobank of dragon IPSCs could also come in handy as a type of insurance policy; IPSCs could be used to treat our dragons should they become sick. For instance, if our dragon had eye problems, we could use dragon IPSCs to make new eye cells to transplant into the damaged or sick eye of the dragon. While such an IPSC-based transplant approach is not yet fully proven to treat human patients, a growing number of clinical trials are underway and there is real hope this might be an effective new form of medicine.[ix]

In Paul's lab, we have made and studied both mouse and human IPSCs over many years. How does this work? We scientists can introduce what are called "reprogramming factors" into ordinary cells (such as skin cells) that turn them into IPSCs. These reprogramming factors are proteins that control the activity levels of specific genes, and together they recode cells such that they "think" they are pluripotent. It works surprisingly well.

IPSCs and ESCs have both also been used to make the reproductive cells of certain animals.[x] Furthermore, IPSCs and ESCs can in principle be used instead of reproductive cells to make a whole new organism, as has recently been achieved for mice[xi] and some other animals, but fortunately not for humans, which we think would be extremely risky and unwise (unlike making a dragon, *right?*)

Neither IPSCs nor ESCs are "perfect" cells as both can acquire mutations over time while being grown in the lab. For these and other reasons, we believe that using human IPSCs and ESCs to make people by reproductive cloning would be a dangerous and unethical thing to do. If you want to read more about these cells and the ethical issues

---

[ix] https://ipscell.com/2017/06/talk-ips-cells-future-genomic-medicine-mtg/

[x] https://www.sciencemag.org/news/2016/10/mouse-egg-cells-made-entirely-lab-give-rise-healthy-offspring

[xi] https://www.nature.com/stemcells/2009/0908/090806/full/stemcells.2009.106.html

that surround them, we'd point you to Paul's book, *Stem Cells: An Insider's Guide* [15].

Unfortunately, dragon ESCs and IPSCs do not currently exist. If they did, we could in theory just use those cells to try to make a dragon, in the ways we've described. Even though we can't make ESCs or IPSCs from an existing dragon now, we could instead try to make (or to obtain) ESCs or IPSC from our starter animal. For instance, we could make ESCs or IPSCs from Komodos, Dracos, or birds.

Do such cells already exist?

Sadly, we couldn't find any evidence that ESCs or IPSCs have, as yet, been made from lizards, but that doesn't mean it's totally impossible or that they don't exist. Maybe someone made such cells, but never published a research paper about them. Still, it might take some effort to get it to work or to track down the necessary cells. Interestingly, however, there are reports that some avian ESCs exist, so that could be another tempting reason to use birds as a starting point to make our dragon [16].

While it would be relatively easier to obtain reproductive cells from our favorite starter animals (although as mentioned earlier, who wants to volunteer to try to harvest sperm or eggs from potentially grouchy Komodo dragons?), having ESCs or IPSCs in hand would offer some other big benefits. ESCs and IPSCs are immortal cells, which means we can grow and expand them in the lab forever, whereas reproductive cells can be tough to obtain and they also do not generally grow in the lab.

We could potentially make billions of ESCs and IPSCs to use in helpful ways in our dragon-building research. For instance, it might prove far easier to carry out CRISPR gene-editing in bird ESCs and IPSCs than in their sperm, eggs or embryos themselves. Then we could use the genetically modified ESCs or IPSCs to make the sperm, eggs, and embryos for reproduction. Still, researchers seem to be getting better at injecting CRISPR into eggs to make gene-edited embryos of various species, so we would probably start with that approach.

No males or sex needed?

As discussed in the previous chapter, we could potentially propagate a group (what we might call a "murder") of dragons from one founding female dragon by parthenogenesis. If producing dragons by parthenogenesis were possible (that is, without needing a male and female dragon to mate), it would become so much easier to expand our dragon population.

One potential serious downside to dragon parthenogenesis though is that all the resulting offspring would be genetically too similar. In turn, this could make our dragons less adaptable as a new species to the world they live in, now and in the future. This could then lead to less healthy dragons and fewer survivors. Ultimately, they might soon become extinct as a species because of a lack of genetic diversity. Our goal is definitely not to make dragons only to have them then soon disappear.

Reproduction by cloning or by some other new method is another alternative to sexual reproduction, but these techniques raise the same problem as parthenogenesis-based reproduction – a lack of genetic diversity. Still, the overlapping research area where stem cells and reproduction meet is evolving so rapidly that you never quite know what is possible.

Just in 2018, researchers reported the generation of mice from same-sex parents [17]. The most successful outcomes came from using two mothers (and no fathers).[xii] While some of the resulting mice weren't healthy, others seemed fine and even had babies of their own later. In contrast, while some embryos generated from two fathers (and no mothers) made it to the end of development in the uterus, that's as far as they got. Only 2 out of 12 survived even 48 hours after birth.

Stanford University law professor and science policy expert, Hank Greely, discusses the possibility of humans reproducing without sex in

---

[xii] https://www.nationalgeographic.com/science/2018/10/news-gene-editing-crispr-mice-stem-cells/

his book *The End of Sex* [18]. One way that sex could become less important for human reproduction is through cloning.

Cloning class

Cloning is likely to play a role in our dragon building project. So how does it work?

As experienced gardeners will know, some plants can be cloned simply by taking a cutting from a plant (little snippets of plant parts) and planting them. The cutting then grows to create a whole new but genetically identical plant. Animals generally cannot propagate themselves in this way.

So how do we clone a whole, new animal? The process begins with an egg, which (like other cells) contains a nucleus. A nucleus is a structure in a cell that contains all its DNA. During fertilization, the DNA of an egg and sperm cell combine to form a new set of DNA. This is packaged into a single nucleus in a one-cell embryo, called a zygote.

In cloning, you bypass that fertilization process. Instead, you remove the egg's nucleus (and hence its DNA) and you never involve sperm. Once you've removed the nucleus from the egg, you use a long, thin needle to carefully insert the nucleus from another cell inside of the egg that no longer has its own nucleus – for example, the nucleus of a skin cell obtained from an adult animal. Now you have a hybrid: an egg with an adult cell nucleus.

Curiously, if you now give that hybrid egg a bit of an electric or chemical shock, sometimes it'll now "think" it's a one-cell embryo and it will then start to develop as if it is a zygote. At the end of in utero development, the trillions of cells in the clone all will have the same DNA that came from the transferred adult cell nucleus. If the cloned embryo continues to grow normally for a few days in a dish in a lab, then it can be implanted (precisely and gently inserted *via* a medical procedure) inside a foster mother's uterus. If it makes it through development and to birth, then that baby will be genetically identical to the individual that provided the adult cell. This is called "reproductive cloning," a process now frequently used in livestock breeding.

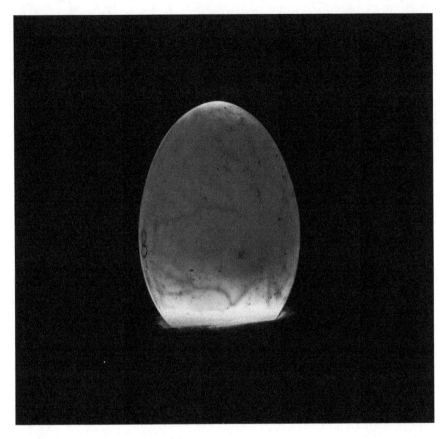

Figure 6.4. A developing chick inside an egg is visualized using "candling." You can make out the blood vessels that help to nourish it during development and the darker "dot" near the middle is likely the chick's developing eye. Image from Shutterstock.

You can imagine that if a farmer had an amazing, one-in-a-million cow, created by normal breeding, there's only a small chance of that farmer creating such a remarkable specimen by normal breeding that cow to a bull, given that sexual reproduction can be such a random process. However, through cloning, you could make another copy – that is, a replicate – of that amazing cow. Sometimes cloning doesn't work so well though so it's not always a sure thing.

In the second kind of cloning called "therapeutic cloning," you go through the same steps, but you do not implant the created embryo. Instead, you use it to make ESCs (as discussed earlier in this chapter).

There are many myths and misunderstandings about cloning. This set of cloning myths and rebuttals published online by the US Food and Drug Administration (FDA) is worth a read.[xiii] It focuses on the cloning of livestock animals and emphasizes how well cloning works to produce normal, healthy offspring. However, it also briefly mentions a developmental problem, called large offspring syndrome (LOS), which is seen in some clones. The affected animals end up growing much too large in a way that is unhealthy.

We have to say that we don't know whether our cloned dragons, or cloned "starter" or intermediate creatures, would be at risk of LOS – or of any other problem caused by cloning – but it's possible. The other issues we'd need to bear in mind are those known to be caused by CRISPR-based gene-editing, such as growth abnormalities, which have been reported in some animals undergoing genetic modification.[xiv]

Another cloning fact mentioned by the FDA also could complicate our plan to use birds to build our dragon. The FDA states that so far it has been impossible to clone birds. It is unclear at this point if anyone has successfully cloned birds since that FDA article came out.

One other possible stumbling block here could be the problem of the bird egg itself. If we could clone a bird embryo, would it just naturally grow its own yolk and shell? Perhaps it would, but if not, then we might need to transfer it into an already existing developing egg that has had its initial embryo removed. Implanting a cloned embryo inside a developing bird's egg (that has had its own embryo removed) could be tricky to do in such a way that it doesn't destroy the egg and allows development to occur in an otherwise natural, healthy fashion.

[xiii] https://www.fda.gov/animalveterinary/safetyhealth/animalcloning/ucm055512.htm

[xiv] https://futurism.com/the-byte/gene-editing-mutated-animals-crispr

Figure 6.5. If we could peer inside a developing dragon egg using a method far better than candling this is what a developing dragon embryo might look like, as created by artist Gareth Monger as a vision of a possible pterosaur embryo. Image copyright of Gareth Monger. Used with permission.

Note that the presence of the developing chicks (and sometimes other birds in their shells too) can be seen while still inside the egg using a process that's been around for ages, called "candling." Due to the semi-transparency of egg shells, egg whites (the developing gel-like substance around embryos), and the chick embryos themselves, shining a bright light on an egg can give reveal some of what is going on inside. With candling, mainly you can see all the blood vessels nourishing the developing chick, but sometimes you can see part of the chick itself including its eyes.

Candling was originally done with candles, hence the name, but is now typically done using bright electric lights. See Figure 6.4 for an example of candling, where you can kind of see the chick, its eye, and the blood vessels that nourish it all still inside the egg.

While the equivalent of "candling" can't be done with pregnant women, ultrasound uses sound waves to achieve something similar, and in fact, far more precise. In Figure 6.5, we can see what a developing dragon embryo might look like, as created by artist Gareth Monger, while still inside its egg.

Cloning, as we've already mentioned, creates genetically identical individuals and this can leave an entire population generated by cloning vulnerable to the same genetic or infectious diseases. Seedless bananas, for example, are usually generated by cloning and are widely at risk from molds.

Like many seedless fruits, these bananas are triploids (meaning they have three copies of every chromosome, whereas normally bananas and we humans have two, one from each parent). These bananas, therefore, cannot reproduce sexually (that's why they have no seeds) and so must be reproduced by cloning. At least plant cloning is far easier than animal cloning. However, all those cloned banana trees are in danger now from the same kind of mold because they are genetically the same.

It would be devastating to create many cloned dragons and then have them wiped out by some disease because each dragon had the same disease susceptibility. The cloned dragons would also all be the same color and size, which might be somewhat boring and it could be hard for us to tell them apart.

Chimeras and chimeric embryos

As discussed in previous chapters, to achieve a dragon with a good balance of various dragon-like features, we might need to make chimeric embryos, and this also could prove useful if cloning runs into trouble. For instance, let's say we want a dragon that is bigger than a Draco lizard and more like the size of a Komodo dragon, but we also want wings based on the Draco's patagia. To achieve both goals, we could fuse embryos (or individual embryonic cells) from a Draco with those of a Komodo.

Because these are somewhat related creatures, such chimeric embryos might develop normally, in some cases with a mix of traits and hopefully those that we want. For instance, you could imagine the resulting chimera being half the size of a Komodo and having big, flappy patagia. It wouldn't be a real dragon or the type that we want yet, but it'd be a big step forward.

Unfortunately, chimera production is a somewhat unpredictable process, and they don't always turn out the way one might hope. Chimeric animals, much like hybrid plants such as that beautiful tomato plant in your garden, also don't generally breed true for the next generation. As a result, our hoped for Komodo-Draco dragon-like chimera could produce babies with a whole range of unexpected traits, but it's unlikely to produce offspring similar to itself. Still, the offspring of this "F1 generation" as we call it in science could have some cool, heritable dragon-like traits, but it'd be a hit-and-miss approach to take if we want to consistently get closer to making dragons.

More likely what we'll do is sort of follow in plant breeders' footsteps, in terms of how they consistently manage to develop amazing plant hybrids. We might, for example, use genetic modification to make a new kind of Draco (let's call it Draco 2.0) with the desired genetic changes hardwired into its genome *via* CRISPR-based gene-editing. We could try the same approach to create a new kind of Komodo (Komodo 2.0) – it could be engineered to have unusually structured forelimbs with long fingers that could be used to generate wings.

One would then breed Draco 2.0 with Komodo 2.0 (or more likely do IVF using their sperm and eggs since the Komodo would likely just eat the Draco rather than try to have sex with it). We'd then hope that the resulting chimeric offspring are healthy and have a range of dragon-like traits inherited from their parents.

The resulting hybrid could be something very dragon-like or it could be a mess. But if we managed to create just one fertile offspring that provides us with a genetic stepping stone towards making a dragon, we could likely produce more from the same parents. This is going to require a lot of trial and error, but at some point, we'd hope to get a modified Draco and a tailor-made Komodo that consistently make hybrid dragon-like babies *via* IVF.

Figure 6.6. The fabled beast of Greek mythology, The Chimera, was part lion, goat, and snake, and it was sometimes described as breathing fire. Image from Shutterstock.

Instead of using Dracos and Komodos, we could also try to make other hybrids or chimeric embryos, such as between birds and lizards, using the same general kind of process.

Whatever creatures we use to make a chimera from, things could go very wrong, and we could end up with a beast more like a mythological monster called The Chimera (Figure 6.6). In *The Iliad*, Homer described The Chimera as a fire-breathing (good from a dragon-building perspective) creature with features of a lion, goat and a back end like a snake (maybe not so good).

CRISPR

On and off in this book, we've mentioned CRISPR-Cas9 gene-editing and how this approach might be useful for dragon building. In this section, we focus more specifically on how we might use CRISPR to generate or accentuate dragon-related traits in our starter creatures.

Figure 6.7. A sketch by Paul that shows how CRISPR-Cas9 can be compared to a Swiss Army Knife-like superhero that edits the genome. From the book *GMO Sapiens*.

What exactly is CRISPR gene-editing and how does it work?

CRISPR-Cas9 is a bacterial weapon that evolved out of the war between bacteria and the viruses that infect them. When viruses infect bacteria, bacteria try to protect themselves by destroying the genomes of these viral invaders. And to do this, bacteria have evolved a whole weapons system that can chew up the viral genome.

However, bacteria don't want to destroy their own genomes or waste energy by turning on their weapon when they don't need to. So they have evolved special proteins, enzymes called nucleases that can chew up DNA or RNA. But these enzymes are highly specific and will recognize only specific stretches of viral DNA.

CRISPR-Cas9 is one such anti-viral system used by bacteria – it can specifically destroy viral DNA.[xv] CRISPR stands for "clustered regularly interspaced short palindromic repeats," which is quite a mouthful to say. These repeats are genomic elements that past generations of bacteria have taken from the viruses that have attacked them. The repeats effectively serve as a memory of past invaders – in a sense, they act as the "GPS" for the nuclease (DNA chopping) part of the system, which is Cas9.

In this way, when a virus attacks a bacterium that has CRISPR-Cas9 (or a similar system), the bacterium can recognize the invader by its genome, which it then chews up using Cas9.

But how did we come to use this bacterial weapon as a beneficial tool for genetic engineering?

Some clever scientists realized that CRISPR-Cas9 could be repurposed and directed to recognize the DNA of any organism. Instead of relying on viral DNA for the CRISPR sequences, you can place any DNA sequence into the system. That same sequence would then be found by CRISPR-Cas9 and snipped, allowing you to in turn make specific genetic changes inside a cell. This adapted gene modifying system should in theory work in the cells of any animal – from lizards,

---

[xv] https://ghr.nlm.nih.gov/primer/genomicresearch/genomeediting

birds, or any other potential starter creature we decide to use in our efforts to make a dragon.

Researchers now use CRISPR-Cas9 to introduce specific changes into DNA, or to correct existing errors (a.k.a. "mutations"), changing them back to the normal sequence. Furthermore, more radical changes are possible. For example, entirely new genes could be inserted into the genome of reproductive cells from another animal (say bird genes going into a reptile embryo) to produce a genetically hybrid animal. We'll talk more about this new gene approach a little later.

In Figure 6.7, Paul has drawn a sketch that portrays CRISPR-Cas9 as a Swiss Army Knife-like superhero.

Using CRISPR to make a dragon

What would the ideal outcome be of using CRISPR in certain steps in the process of trying to make a dragon?

We could use CRISPR to genetically modify reproductive cells or pluripotent stem cells obtained or created, from our starter animal. If things go according to plan, the resulting new creature will possess the intended genetic modification and no other (we don't want CRISPR to make additional, unwanted DNA changes called "off-target activities," see more on this below). Ideally, these introduced genetic changes will give our creature dragon-like attributes, but no other undesired traits. For instance, we'd want a dragon with two wings, not three or four. We want a dragon that breathes fire out of its mouth and not its butt. You get the picture.

So how would this work?

We'd need to introduce the desired genetic change very early in our starter animal's development, for the reasons we've explained earlier in this chapter. This means introducing it into reproductive cells, pluripotent stem cells or a single-cell embryo. During development, one embryonic cell becomes two, then four, and soon you have the trillions of cells all genetically identical, or at least close to it (some random mutations can rarely pop up during all those cell divisions since DNA replication isn't entirely perfect). In this way, the specific, desired genetic change made by CRISPR in the starting embryo will be present in all the final dragon's cells.

The specific CRISPR-induced genetic changes we'd introduce could include many of the kinds of things already discussed in this book. For instance, we might change genes associated with wing formation or with patagia development in a Draco lizard to make its patagia more like wings. Alternatively, we might start with a bird and change genes that will eventually enable it to breathe fire. The exact genes for fire-breathing aren't known, of course, so that presents a big challenge, and we would have to do quite a few experiments along the way.

Another challenge in terms of CRISPR'ing animals is that we might not yet know the entire genomic sequence of some of our favorite starter creatures. For instance, as far as we can tell the Draco lizard has not had its genome entirely sequenced, so we might have to do that first ourselves. You need to know an animal's genome sequence, or at least the sequence of the genes you want to target with CRISPR, before you start trying to change them. Even the "same" gene has some different DNA bases in various animals. Fortunately, the Komodo genome was just sequenced.[xvi]

It's also good news that many bird genomes have already been sequenced.[xvii] And a final bit of additional good news is that the cost of DNA sequencing whole genomes has become more affordable in recent years, which would allow us to sequence our starter animals ourselves.

In addition to changing already existing genes in our starter animals, we could use genetic engineering technology to introduce entirely new genes into them, as we aim to achieve specific traits. For complex or even non-existent traits, like fire-breathing, it might not be enough to tweak existing bird or reptile genes. Entirely new ones (e.g. from a bombardier beetle) might be needed to get the desired result.

But using CRISPR to genetically modify the genomes of animals is not without its risks.

If you want to know more about the issues involved, particularly with use of CRISPR in humans, Paul wrote a book on this topic a

---

[xvi] https://www.biorxiv.org/content/10.1101/551978v1

[xvii] http://science.sciencemag.org/content/346/6215/1311.full

couple of years ago, called *GMO Sapiens*. It discusses the technologies involved in hypothetically making genetically modified human beings, including using CRISPR gene-editing. In that book, Paul discusses in detail what could go wrong if we go down the path of human genetic engineering, including designer babies and eugenics.

Many of the same kind of technologies involved in trying to make genetically modified people (for example, people who could not be infected by various viruses) would also come into play if we use CRISPR in our dragon-making project.

What could go wrong and how we might die?

All kinds of things might go awry when it comes to genetically engineering animals to make dragons. One of the most likely problems is that – instead of making dragons or dragon-like creatures – our genetic tinkering creates weird, unpredictable creatures. There are other risks too, such as the production of various hybrid creatures that could be even more dangerous than full-blown dragons.

There are also many ethical issues to consider when it comes to this and other possible kinds of negative outcomes (more on that in Chapter 8).

Dragon building technology

Hopefully, at this point you have a good sense of the potential ways in which cutting-edge technologies such as CRISPR, stem cells, and reproductive manipulations could help us to try to make a dragon. At the same time, you probably have a clear feel for how risky this would be.

References

1.   Knoepfler, P., *GMO Sapiens : The Life-Changing Science of Designer Babies*. 2015, World Scientific Publishing, Singapore.

2. Curtis, N.R., Firefly encyclopedia of reptiles and amphibians. *Library Journal* 2003. **128**(1): 88–88.

3. Li, Y., *et al.*, 3D printing human induced pluripotent stem cells with novel hydroxypropyl chitin bioink: scalable expansion and uniform aggregation. *Biofabrication* 2018. **10**(4): 044101.

4. Pascoal, J.F., *et al.*, Three-dimensional cell-based microarrays: printing pluripotent stem cells into 3D microenvironments. *Methods Mol Biol* 2018. **1771**: 69–81.

5. Hockman, D., *et al.*, The role of early development in mammalian limb diversification: a descriptive comparison of early limb development between the Natal long-fingered bat (*Miniopterus natalensis*) and the mouse (*Mus musculus*). *Dev Dyn* 2009. **238**(4): 965–979.

6. Nolte, M.J., *et al.*, Embryonic staging system for the Black Mastiff Bat, *Molossus rufus* (Molossidae), correlated with structure-function relationships in the adult. *Anat Rec (Hoboken)* 2009. **292**(2): 155–168, spc 1.

7. Hockman, D., *et al.*, A second wave of Sonic hedgehog expression during the development of the bat limb. *Proc Natl Acad Sci USA* 2008. **105**(44): 16982–16987.

8. Perez-Rivero, J.J., A.R. Lozada-Gallegos, and J.A. Herrera-Barragan, Surgical extraction of viable hen (*Gallus gallus domesticus*) follicles for *in vitro* fertilization. *J Avian Med Surg* 2018. **32**(1): 13–18.

9. Li, B.C., *et al.*, The influencing factor of *in vitro* fertilization and embryonic transfer in the domestic fowl (*Gallus domesticus*). *Reprod Domest Anim* 2013. **48**(3): 368–372.

10. Nomura, T., *et al.*, Genetic manipulation of reptilian embryos: toward an understanding of cortical development and evolution. *Front Neurosci* 2015. **9**: 45.

11. Bogliotti, Y.S., *et al.*, Efficient derivation of stable primed pluripotent embryonic stem cells from bovine blastocysts. *Proc Natl Acad Sci USA* 2018. **115**(9): 2090–2095.

12. Tachibana, M., *et al.*, Human embryonic stem cells derived by somatic cell nuclear transfer. *Cell* 2013. **153**(6): 1228–1238.

13. Chung, Y.G., *et al.*, Human somatic cell nuclear transfer using adult cells. *Cell Stem Cell* 2014. **14**(6): 777–780.

14. Inoue, H., *et al.*, iPS cells: a game changer for future medicine. *EMBO J* 2014. **33**(5): 409–417.

15. Knoepfler, P., *Stem Cells : An Insider's Guide*. 2013, World Scientific Publishing, Singapore.

16.   Pain, B., *et al.*, Long-term *in vitro* culture and characterisation of avian embryonic stem cells with multiple morphogenetic potentialities. *Development* 1996. **122**(8): 2339–2348.

17.   Li, Z.K., *et al.*, Generation of bimaternal and bipaternal mice from hypomethylated haploid ESCs with imprinting region deletions. *Cell Stem Cell* 2018. **23**(5): 665–676 e4.

18.   Greely, H.T., *The End of Sex and the Future of Human Reproduction*. 2016, Cambridge, Massachusetts: Harvard University Press.

Chapter 7

After a dragon: building unicorns and other mythical creatures

Our next challenge: unicorns and other mythical creatures

We figure that if we can build a dragon, we can likely construct other interesting, mythical creatures as well. We'd even use some of the same technologies, such as the stem cell and CRISPR-based gene-editing techniques that we discussed in the previous chapter.

Some specific "brought-to-life" mythical creatures are unlikely to be particularly controversial – and might even become cultural sensations like unicorns – while others could raise some fierce debate and troubling ethical issues. For example, many mythological creatures are based on humans or are somewhat human-like, which would make them highly controversial to produce. The new creatures could also create real problems both for society, ecosystems, and for us as their makers.

In our view, building partly human mythological creatures wouldn't be an ethical thing to do. We're thinking of creatures such as elves, centaurs, the Egyptian Sphinx, and bigfoot/yeti to name just four. So, in this chapter, we discuss the idea of building only one partly human type of mythical creature, merpeople (mermaids and mermen), in any depth and just as an educational exercise. But for most of this chapter, we focus primarily on how we could go about making a few particularly cool, non-human mythological creatures.

In each case, we discuss the ways in which we or others might make some of these creatures. The production of each one is likely to begin with a particular "starter" creature (just as we discussed starter creatures for dragon building), the choice of which will depend on the nature and attributes of the intended mythical beast.

We should also note that quite a few mythological creatures that aren't explicitly called "dragons" nonetheless bear a striking resemblance to dragons and have certain dragon-like features. For instance, as already discussed earlier in this book, both The Hydra and The Chimera of mythology were viewed by some believers as actual dragons and each creature had dragon-related characteristics.

So too did the Basilisk, which was a monstrous serpent that could kill plants, animals, and people with either its poisonous breath or with just an evil look. The Cockatrice was even more dragon-like than the Basilisk. Since these beasts are so similar to dragons, they could potentially be made in much the same way as we've described for making dragons. For this reason, although they are cool, we won't discuss them further in this chapter.

We aren't the only ones envisioning the potential production of totally new creatures by using cutting edge technologies, such as CRISPR. Professors Hank Greely and Alta Charo wrote a whole article [1] about this possibility back in 2015, including this particular quote that stood out to us as we were writing this chapter:

> "Considering, as Heinlein[i] did, what humans have done in the last 10,000 years to wolves and their descendants — and continue to do, as in the labradoodle — why should we not expect dwarf elephants, giant guinea pigs, or genetically tamed tigers? Or — dare we wonder — the billionaire who decides to give his 12-year-old daughter a real unicorn for her birthday?"

So where shall we start?

---

[i] Robert Heinlein was a prolific American science fiction writer whose work predicted many future technology innovations including the cell phone.

With the unicorn of course – a popular favorite where mythological creatures are concerned. And we are focusing first on a unicorn because practically speaking we believe that making one is probably not going to be as difficult as making some other mythological creatures.

Unicorns

A brief history of unicorns

Most people are familiar with unicorns, but in case you aren't, they are mythical horse-like creatures that famously have a single, beautiful, and straight (yet spiraled) horn on their head. The name "unicorn" even means "one horn."

In mythological stories, unicorns weren't just pretty and unusual, but were also thought to possess magical powers. They could also be tough fighters should the need arise. Weirdly, it was also believed in ancient times that only human virgins could tame unicorns (Figure 7.1).

The earliest unicorns in mythology or religion might have been based on ancient animals called aurochs, a type of horned wild cattle depicted by the Harappan culture – a Bronze Age civilization that occupied the northwestern regions of South Asia. Unicorns might also have been based on an antelope, called the oryx.[ii] For whatever reason, aurochs were reportedly sometimes referred to as unicorns, including in The Bible.

It's also possible that the idea of a unicorn was conceived by someone who saw an animal that would normally have two horns, but in this instance only had one. Perhaps this single-horned animal had somehow failed to develop normally as an embryo. Or perhaps it had been born with two horns but had lost one due to injury or in a fight, and then was mistakenly thought to be a unicorn by (certain imaginative) people. It has even been proposed that in ancient times two-horned animals seen from a long distance in profile or as a

---

[ii] https://www.wired.com/2015/02/fantastically-wrong-unicorn/

silhouette might have been mistaken to be unicorns with only one horn.

Figure 7.1. Virgin and Unicorn, a painting by Domenichino (1581–1641). Image in public domain.

Some Greeks were convinced that unicorns roamed around what is now the country of India. Ctesias, a Greek physician and historian who had spent time in Persia (modern day Iran) wrote that he had seen an oryx or "wild ass (donkey type creature)" that was somewhat similar to a unicorn.[iii] In Europe, the unicorn is sometimes associated with the Virgin Mary. Unicorns appear in many pieces of medieval artwork, again strangely sometimes

---

[iii] http://content.time.com/time/health/article/0,8599,1814227,00.html

depicted with virgins (other than Mary), perhaps because unicorns were seen as being "pure."

A unicorn's horn was thought to be made of alicorn, a magical substance that could cure many diseases. In the olden days, it was said that unscrupulous traders would find horns from dead land animals or the tusks of dead narwhals (which have a single, long tusk that looks rather like a unicorn horn) washed up on the beach and sell them as alicorn. It was highly profitable given the magic attributable to unicorns.

The Danish physician and naturalist, Ole Worm, finally debunked this practice in 1638. Isn't Worm a great name for a naturalist? He reportedly found the remains of a dead narwhal with the horn still attached to its skull. Based on this discovery, he identified narwhals as the real source of the supposed "unicorn horns." Apparently, he then concluded that there were no real unicorns, perhaps something he had long suspected.

Figure 7.2. Japan Osaka Kaiyukan Sea Aquarium on 25 November 2014 Narwhal. Image from Shutterstock.

Worm also created one of the earliest natural history museums in the world, which was called something along the lines of "Museum Wormanium." Maybe we should make a modern type of museum like that called Museum Knoepflerium? If we are successful with our plans in this book our natural history museum could include living dragons.

Although Worm disproved the existence of unicorns, he didn't entirely escape the mystique associated with them. In one article about him, it was said that he still, "couldn't help but be carried along by the notion that unicorn horns contained an antidote to poison, and with this in mind, he carried out primitive experiments where he poisoned pets and then served them ground-up narwhal tusk. He reported that they did recover, suggesting that this poisoning wasn't quite so efficient." [iv]

The Unicorn is also a symbol of Scotland, as it is their national animal. It is thought by some that Scotland took this mythological animal as its national symbol because, at the time, unicorns and lions were believed to be natural enemies. And lions are the symbol of England. So, unicorns are perhaps a metaphor for the historical conflicts between England and Scotland.

Wired Magazine has also published a funny article on the wacky history of the unicorn. [v] It includes a quote from even further back in time – back to the Roman historian, Pliny The Elder. Pliny made the unicorn out to be more like a bad-ass chimera than a pretty, friendly, and force-for-good unicorn that we might imagine today:

> "The unicorn," Pliny wrote, "is the fiercest animal, and it is said that it is impossible to capture one alive. It has the body of a horse, the head of a stag, the feet of an elephant, the tail of a boar, and a single black horn three feet long in the middle of its forehead. Its cry is a deep bellow."

[iv] http://cphpost.dk/history/ole-worm-the-man-who-studied-unicorns.html

[v] https://www.wired.com/2015/02/fantastically-wrong-unicorn/

We wouldn't want to meet Pliny's unicorn in a dark alley or out on our own in a forest. Would you?

How would we create a unicorn?

It's a good question. It could be quite easy to build a unicorn, at least in comparison to building a dragon, which needs flight, fire, and a reptilian appearance. By contrast, making a unicorn might not be so hard, especially if we don't need to endow it with magical powers because how would we even do that, right?

Figure 7.3. The skull of a black rhino. Note, how there are no bones where the two horns should be. This is because rhino horns almost entirely consist of modified skin rather than having a core of bone. Image source: Vernon Swanepoel, Creative Commons. Image not modified.

One way to go about making a unicorn is to start with a certain species of horse. We could then genetically modify this horse to endow it with a single, long horn that grows out of the middle of its forehead.

At this point, it is time to answer the question raised earlier in Chapter 5, "What is a horn?" The answer partly depends on the specific creature that has the horn. A standard "horn," for example on a mammal – such as an antelope or deer – is basically a long, bony protrusion that emerges from its head and is covered by specialized skin-like layers.

One way to think about giving a creature a new horn is to imagine taking one of your fingers (but with no muscles, tendons, or joints) and simply sticking it to your forehead. Once there, it would fuse with the existing tissue and with the underlying bone of your skull. Not pretty perhaps but a close approximation to a real horn.

Now if you were mysteriously able to simply "slap" a bone onto your forehead and have it merge with your skull, as an adult you'd have some adjusting to do. The new horn would interrupt your field of vision – it would be distracting at least and maddening at worst. If the new horn was several times bigger than a finger in length and mass, your head and body would also have to get used to that new "thing" being there, and to the different weight of your head, and more. By contrast, animals born with horns (such as rhinos) or those that develop them only later as they mature, gradually get used to being horned.

Of course, a typical mammalian horn is not just made of bone. The skin-like layers that cover the bone inside a horn consist mostly of a protein called keratin. You may remember keratin and keratinocytes (the cells that make keratin) from Chapter 5.

The horns of mammals and most other creatures have this kind of architecture, but some animals don't have living bone inside their horns. When we picture a horned creature, many of us will think first of a rhino. But rhino "horns" aren't in a sense typical horns – they don't have bone inside of them but are instead made almost entirely from dense skin-like material. Note how the skull of a black rhino shown in Figure 7.3 looks impressive and has plenty of bone where its nose was, but lacks any bone where the horns were on top.

Even though we ourselves imagine a triceratops dinosaur to look kind of like a modified rhino, it appears that – unlike rhinos – triceratops had true horns with a bony core. Today's horned lizards also have bone-based horns (which we'll keep in mind for our dragon building as well).

So, to make a horse into a unicorn we'd need to genetically modify it to either give it a true horn – one with bone inside – or a horn that's more like that of a rhino. Either could possibly give it the desired unicorn-like appearance and a weapon to fight with, should it need it. However, for an animal with a long, relatively thin horn like a unicorn, bone-based horns would work better.

A now extinct Siberian type of rhino (*Elasmotherium sibiricum*) had one dramatic horn and it is sometimes referred to as the Siberian Unicorn. It is now thought – based on new research – that these "unicorns" might have coexisted with ancient humans, perhaps contributing to early unicorn legends.[vi]

What information is available to help us genetically modify a horse to make a unicorn? What genes might we need to alter? Several genes have been implicated in horn development in some animals, including cattle, and other genes may inhibit horn growth. Some of these candidate "horn inhibitor" genes make proteins that help cells bind each other, while others regulate the movement of keratinocytes.[vii] In cattle, more "pro-horn" genes make proteins that stimulate cell growth, which makes sense as horn growth requires more cells in a particular area as opposed to just an area of flat skin alone.[viii]

Some researchers have been using this kind of information for years to try to breed dairy cattle that don't have horns, which would be very useful for farmers as currently most cattle that have horns must have them removed. This entails a painful, costly procedure called "polling" that is performed on more than 70% of dairy cattle today in some countries to protect their handlers and other animals. But in 2016, groundbreaking research on horn development was combined with gene-editing approaches to make hornless cattle.[ix,x] Based on this

---

[vi] https://thescipub.com/PDF/ajassp.2016.189.199.pdf

[vii] https://www.ncbi.nlm.nih.gov/pmc/articles/PMC3017764/

[viii] https://journals.plos.org/plosone/article?id=10.1371/journal.pone.0202978

[ix] https://www.nature.com/articles/nbt.3560

[x] http://www.sciencemag.org/news/2016/05/gene-edited-cattle-produce-no-horns

discovery, in the future one can imagine far fewer cattle will be born with horns and need to have them removed.

But of course, we don't want to remove horns. We want to produce horns in horses to make them into unicorns. And we could use what we learn from making a unicorn to endow our dragon with various kinds of horns as well.

It is reasonably plausible to modify gene activity in a horse in a specific way that gives it a single horn. However, there's another unicorn characteristic that requires our attention, which could prove more difficult to attain. Unicorn horns are supposed to be straight, but almost all "horned" animals have curved horns.

In theory, we could emulate the horn of the narwhal (whose proper name, *Monodon monoceros*, apparently means "one-horned unicorn" and note amusingly that "rhinoceros" means "nose horn") with its single, beautiful "horn," which is actually a tusk.

If we went this route, we could aim to generate a similar tusk in a horse (Figure 7.2). However, since the narwhal's tusk goes through its head to jut out into the world, we'd have to deal with that unfortunate reality in our horse. We're just not sure how to safely make a horse with a giant tusk projecting from its head. For this reason, sticking with actual horns (rather than going with a giant tooth) seems like a wiser bet.

Note that narwhals are most closely related to beluga whales. While the Latin name for narwhals, *Monodon monoceros*, has a straightforward meaning as mentioned earlier, in contrast the name "narwhal" itself has more complex, even dark, origins. It most likely derives from the Icelandic term for "corpse whale," perhaps reflecting the terror felt in sailors when first seeing the stark, white, corpse-like color of narwhals.[xi] The narwhals "horn" could have freaked out the sailors too.

We weren't able to find much information on narwhal tusks, but notably these unusual teeth, while projecting straight out, actually spiral in a specific helical way up their axes.[xii] Many assume that a narwhal

---

[xi] https://www.britannica.com/animal/narwhal

[xii] https://www.ncbi.nlm.nih.gov/pubmed/30263923

uses its tusk only for defense, hunting, or in mating rituals, but it actually appears to also be a unique sensory organ, which could be an interesting addition to our dragon. However, exactly what the narwhal tusk senses remains to be discovered.

Figure 7.4. A statue of "Pegasus being held back by Renown" by artist Eugène Lequesne in Paris. Public domain image by Marie-Lan Nguyen.

Taking wing with Pegasus

A brief history of Pegasus

Pegasus is also in the pantheon of horse-like mythical creatures. In Greek mythology, Pegasus was the winged horse of the demigods and heroes, Bellerophon and Perseus. Pegasus was reportedly born from the blood of the Gorgon Medusa (when she was slain by Perseus) and from the ocean (its cresting waves were thought to have created horses.) Perseus used Pegasus to rescue the maiden Andromeda from a sea monster. The hero Bellerophon captured Pegasus with a golden bridle after praying in Athena's temple. He was famed in mythology for slaying mythical monsters while riding Pegasus, including the fearful dragon-like Chimera (remember chimeras or combination creatures?), the Amazons, and various pirates. A horse that can take to the skies would be a powerful creature to have.

How would we create a flying horse like Pegasus?

What if we actually wanted to make winged horses like Pegasus? Starting with a horse, we could go through some of the same processes we discussed back in Chapter 2 to give our dragon the ability to fly.

However, as big as Komodo dragons are compared to say Draco lizards, horses are much larger animals than Komodos. The average adult horse might weigh five times as much as a Komodo. And you can likely recall from earlier in the book that the more an animal weighs, the harder it is for it to leave the ground and fly. Most horses are so heavy that they would need a wingspan of perhaps 40 feet (ca. 12 meters) to get aloft, which might not be compatible with an actual living creature that stays intact during flight.

For these reasons, right off the bat, we would need to start with a very small horse – or even better, a smallish pony. Unfortunately, in most depictions of Pegasus it has wings that are far too short to get it airborne. If you look at the sculpture of Pegasus in Figure 7.4, it has impressive wings but compared to its huge body size, they aren't going to get Pegasus off the ground, barring some magic. And in this book, we aren't relying on magic.

Another challenge is that horses already have four legs. You might also recall from Chapter 5 that the default developmental plan for vertebrate animals is four appendages, not the six that would be needed for a creature to have four legs plus two wings. Still, many dragons are depicted as having four legs plus two wings, but they'd be harder to make than the wyvern version of a dragon, which only has two wings and two legs.

Researchers have produced flies with extra pairs of wings (four total instead of two) by modifying certain genes so sparking wing development on the back of a horse could be possible.[xiii] The scientists made four-winged flies by duplicating the whole thorax (the region that wings grow out of in flies) though so doing the equivalent in a horse could produce a clunky creature.

Despite these challenges, it might be possible to make a winged horse. And although it might be technically difficult, what with its large size and weight, at least we wouldn't have to make it breathe fire.

Hippogriffs and Griffins

A brief history of hippogriffs and griffins

Some mythological, creatures were quite similar to each other, almost like different versions of the same thing. Hippogriffs and griffins are such an example, as are the many dragon-like creatures we've discussed.

While today, most people might associate hippogriffs with the Harry Potter books and movies, the author of the Harry Potter books, J.K. Rowling, did not invent them. They actually come from Roman or even earlier times. Hippogriffs are hybrids (or chimeras) – their bottom half is that of a horse while their top half is an eagle. These powerful creatures were depicted as being able to both fly and run – very fast! You wouldn't want to mess with them.

[xiii] https://www.mun.ca/biology/scarr/Bithorax_Drosophila.html

Figure 7.5. A depiction of a griffin in a Medieval tapestry, Basel, c. 1450. Artist and photographer unknown. Public domain image.

Griffins were thought to be the fathers of hippogriffs (hence the "griff" in its name). They were also hybrids but in this case of lions (the back half) and eagles (up front).

Griffins seem to predate hippogriffs, and are present in both Greek and Roman mythology, as well as in the art from other ancient civilizations, including from Egypt. You can see an example of a griffin, depicted in a medieval tapestry in Figure 7.5. There are all kinds of interesting stories involving griffins, including the idea that their nests (they laid eggs) contained gold nuggets. Depending on the artist and civilization, griffins were sometimes depicted as being more lion than eagle, or *vice versa*.

Griffins are often used as symbols. While many mythological hybrid creatures have been perceived to be monsters or evil, Griffins it seems were viewed more favorably. At times, they were even used as religious or other symbols. In fact, griffins are the symbol of Paul's undergraduate school, Reed College.

How to build a hippogriff or griffin?

Hippogriffs were believed to be produced when a griffin mated with a female horse (a mare). For this reason, we think that it would make the most sense to make a griffin first. If we are successful, we could then mate our griffin with a mare to generate a hippogriff (because we believe in mythology, right?)

As we've said, griffins are part eagle and part lion. And by now, you should hopefully be thinking "chimera" given what you've already read in this book. So how would we make such a chimera? One way would be to fuse early embryos of lions and eagles together. However, biologically eagles and large cats, such as lions, are not at all closely related to each other. This makes it highly unlikely that such a chimera would survive even just through embryonic development. Still, it's at least hypothetically possible.

Another factor to consider is that griffins were believed to lay eggs. So, we'd somehow need to manipulate an eagle's egg such that we can remove the top half of the eagle embryo and then fuse it to the bottom half of a lion embryo, all without irreparably damaging the embryos, the egg or other reproductive structures. Alternatively, you could mix the cells of some newly formed eagle and lion embryos (for example, using cells from embryos that are just a few days old; younger embryos would be better) and hope for the best. But we think it'd be a long shot.

Another challenge is that as much as eagles are giants amongst birds, they are way smaller than the average lion. Imagine a griffin with a normal lion-sized back end and a much smaller eagle upper body or head. That isn't going to work. This disparity means we'd either have to shrink the lion or boost the growth of the eagle half. We'd probably aim for the latter if we were to pursue this project.

A tail of Mermaids and Mermen

A brief history of merpeople

Mermaids and more broadly "merpeople" appear surprisingly often in separate myths throughout the world, suggesting a common human fascination with the idea of creatures who are part human and part fish. In Mesopotamian and Assyrian cultures, the goddess Atargatis became a mermaid. Apparently, some believed that Alexander the Great's sister Thessalonike turned into a mermaid, after asking if her brother was dead and getting an unfavorable answer. And in the tale *1001 Arabian Nights*, sea people (and specifically a girl named Djullanar) helped heroes, such as Abdullah and Bulukiya, breathe underwater.

In Western Europe, mermaids were usually worrisome omens whose appearance foretold bad weather for ships and sailors. The Selkies of Scotland and Ireland were seal women who were similar to mermaids. They could go back and forth between a fully seal-like and human state. At some point, they would shed their skin and marry a human husband. If he found her skin, which she had hidden in a trunk, she would have to stay on land with him.

In Greek culture, sirens (who were another chimera – this time half bird, half woman) would sing to sailors and lead them to a watery death by causing them to crash their ships onto rocks.

Later in history, sirens were depicted as being more fish-like than bird-like, with the integration of other mermaid type creatures from different cultures. Today, sirens are typically considered and depicted as a type of mermaid.

The Chinese epic called the Shanhaijing, which dates back to the 4th century BCE, has numerous mentions of sailors sighting mermaids, who cry pearls as tears.

Figure 7.5. *The Little Mermaid* statue by Edvard Eriksen, which sits on a rock in Copenhagen's harbor. Creative Commons image, not modified.

In Japanese folklore, there are the Ningyo, which roughly translates as "human fish." They sometimes were depicted in drawings as mermaid-like creatures. Ningyo were thought to have special powers and if a human ate a Ningyo they might live for hundreds of years.[xiv]

The Philippine creature, the Siyokoy, is also a mer-like creature, and at one time, the Javanese people believed in a mer-queen named Nyi Roro Kidul. In the ancient Indian (Cambodian and Thai) epic poem, Ramayana, a mermaid princess called Suvannamaccha falls in love with a Monkey God named Hanuman.

Mermaids are present in myths from around the world. Other examples include Russian mermaids or Rusalkas, which were depicted as the restless souls of undead women, who had usually drowned. In the Caribbean, there are depictions of merfolk that are inspired by the goddesses Jagua and Yemoja.

Perhaps the mass popularization of mermaids (especially mermaids with a positive connotation) comes in part from the Hans Christian Andersen tale of the Little Mermaid, and its later adaptation to animated film by Disney. The fairy tale was much darker than the Disney movie suggests. In the original Danish fairy tale, the Mermaid must kill the prince after he has chosen to marry another, because his blood will return her feet to fins. But she can't bring herself to do it, and throws herself off a ship and drowns. Perhaps this would have been too dark a tale for a children's movie, but it does show us how mermaids, until recently, were thought of by many cultures as being ominous.

There is also the 20th-century superhero Aquaman, who has some merpeople-like powers even if he doesn't quite look the part. Although possibly one of the oddest modern-day superheroes, the 2018 *Aquaman* movie has proven to be quite popular with some audiences.

[xiv] https://japanesemythology.wordpress.com/toyota-mahime/ningyo/

Hypothetically, how could we create a mermaid (not that we would ever try)?

There would be both technical and ethical dilemmas in creating mermaids. On the technical side, we'd most likely need to make a human-fish chimera to generate a mermaid. While, as we've discussed earlier in this book, it is possible to make chimeras by combining parts of the embryos of different creatures, usually those creatures are evolutionarily related or similar to each other. Humans and fish are not close at all on the evolutionary family tree of life, which means that a human-fish chimeric embryo would be unlikely to survive.

Alternatively, instead of going the chimera route, we could start with human development and try to make the resulting human more like a fish by making some specific genetic changes. Again, this would be unethical to even try. But playing it through in our heads, we could give them scales (a kind of modified, scaly skin), gills to breathe with while underwater, and fins (which are not so different to patagia in a way). And finally, maybe we could fuse their legs together to create the tail flipper that is typical of merpeople. In fact, some merpeople are much more similar to humans than they are to fish. For instance, the famous *Little Mermaid* statue in Copenhagen, by Edvard Eriksen, does not have a fish tail. Instead, she has two fused human legs that each have fins attached to their end (see Figure 7.5). A lab-generated mermaid similar to *The Little Mermaid* would be less difficult to produce than those whose lower body is truly half fish.

Why it would be unethical to create a Mermaid

Beyond the technical challenges, there are some serious ethical issues with such a project. These issues are too grave to ever consider proceeding with it. Any kind of human chimera research, even that restricted just to the lab (let alone trying to make an actual living human chimera) is controversial.

We discuss this kind of thorny ethical issue in the next chapter, but just a few of the concerns that might come up include the potential creation of unintended part-human creatures (not just merpeople but other part-human animals might arise too), and the controversy over making a mixed species creation that is part human.

And to make such a creature that is half-human and half-fish would be profoundly unethical. For one thing, the merperson who'd be created by such research would have no ability to consent to it. They could also develop all kinds of health problems. Even if they remained healthy, they might face negativity from some people, although we suppose they could end up being celebrities who are both happy and healthy.

We'd run into similar ethical minefields with the creation of other part- human mythical creatures as well. In Greek mythology, centaurs were part horse-part man and were famously good archers. Chiron is perhaps the most famous centaur, and he was known as the greatest "hero trainer" in Greece (he trained the Greek warrior, Achilles). He is also associated with the Greek star sign, Sagittarius.

Could someone try to make a real-life Chiron?

While humans and horses are likely more compatible as chimeras than humans and fish are, this is another clear "no-go" for ethical reasons.

Make myths real

After thinking through all these mythological creatures, if we were going to make a real version of another kind of mythological animal beyond a dragon, we would start (and maybe stop) with a unicorn. The other kinds of creatures discussed in this chapter present all kinds of technical and ethical roadblocks. It's still interesting to think about how technology could turn myths into reality.

A unicorn would be an exciting creation like a dragon, but far safer to us and the world. In fact, we wouldn't be shocked if someone actually tried to make a real unicorn in the next 10 years.

Reference

1.   Charo, R.A. and H.T. Greely, CRISPR Critters and CRISPR Cracks. *Am J Bioeth* 2015. **15**(12): 11–17.

Chapter 8

The ethics and future of engineering dragons and other new beasts

Dragon ethics

Part of being a professional scientist is that one willingly agrees to do their research within certain ethical parameters and guidelines. This chapter is an exploration of the ethical constraints on cutting-edge research in general and more specifically on trying to build a dragon.

So, let's imagine for the moment that we did it! We successfully built a dragon based on all the planning and research that we've discussed in the previous chapters.

But *should* we have done it?

What kind of trouble did we get into along the way? Are we even still alive? Have other people been hurt? And what about the animals we used in our research to make our dragons? Furthermore, was all of this a good thing for the dragon or dragons we built?

So far in this book, we've focused mainly on *how* to go about making a dragon, but in this chapter, we address whether we *should* have and the ethical issues our actions would have raised. As just mentioned in the previous chapter, making other new creatures would raise many of the same difficult issues. As Jeff Goldblum's character says in the movie *Jurassic Park*, "Your scientists were so preoccupied with whether or not they could, that they didn't stop to think if they should." The plot of the movie would generally answer that question, "No, you shouldn't have made dinosaurs!" But, once they had done it, it was too late.

And of course if the *Jurassic Park* scientists had said to themselves, "You know what, we can't build dinosaurs because it would be potentially too dangerous or unethical. What were we thinking? Stop the project!" then there would have been no dinosaurs (or at least, no pretend dinosaurs in the movie). The movie would also have been a box-office dud.

There's a similar situation here with our book. If we had at the outset just said, "Nope, it's too dangerous and ethically complicated to make a dragon" then this book wouldn't have been too interesting. We likely wouldn't have written it at all.

And also, we're hoping that you read the Preface and paid attention to the book's subtitle, "*A satirical look at cutting edge science.*" This book is indeed meant to be taken with a large pinch of salt – actually, we are satirizing some of the hype that surrounds cutting edge science.

Still, with all of this in mind, let's take some time here to ask some specific, tough, but appropriate ethical questions before we make a dragon or even a group or "murder" of dragons.

Should we even make a dragon, or would it be an inherently unethical thing to do? What are the specific ethical issues involved? Is the process itself too dangerous to ourselves and to others to even try? And if we are successful, what new dangers and ethical issues arise then as well?

We ought to answer some of these questions and address the tough dilemmas they raise *now* rather than later when we already have a dragon. History tells us that once you've done something radical in science, it's often too late to consider the ethics after the fact.

You can't put the technological genie back in the bottle.

Too dangerous for humanity?

The dangers associated with trying to make a dragon – and then all the dangers that will arise once we have introduced one or more dragons into the world – warrant some serious discussion.

You've perhaps realized from the countless warnings littered throughout this book that building a dragon would be extremely dangerous. For example, there's a real danger that hundreds of people, especially those of us right in the immediate "blast zone," could die or be seriously injured. That's more than a tad worrisome. Even if all goes well in the production phase, there's no guarantee that our dragon won't go on some crazy murdering spree.

In reality, our dragon could cause the deaths of hundreds of people – out of hunger, instinct, or for its own entertainment. And if we try to control the dragon's every move (which is probably impossible anyway) what we'd end up creating is a puppet rather than a sentient, living creature. If our dragon is highly intelligent, both intellectually and emotionally, then constantly controlling it wouldn't be right. It could easily rebel against us if we were too heavy-handed.

And what if our dragon behaves and is happy? (And how would we even know if it's happy unless, of course, it can talk, as we hope it will?) Our intentions for building it are still somewhat selfish. We would like the dragon to feel like it is part of the family and not just a "showpiece" but, in the end, that's exactly what it could be if we don't try to make our dragon much more than that.

This all requires some deep contemplation.

Are there any solutions to these concerns?

While no possible solutions are guaranteed (obviously), perhaps the best idea we have of keeping ourselves and others safe from our dragon is that of engineering an "off switch," as we discussed in Chapter 5.

Humans as friends or family, not food

We want and need our dragon to see humans as friends, not food. One way to achieve this is to ensure that our dragon interacts with people from an early age, while it's still capable of causing minimal damage. Our hope is that it will learn to be comfortable and caring around humans who treat it positively and even lovingly before it grows into a dangerous beast. And we plan to control our dragon's food, interactions, and atmosphere for the first years of its life to instill good habits and health.

Hopefully then, when it's older, it won't have to be constantly watched, although constant supervision might be necessary with a dragon even under the best of circumstances. Rather than having a "domesticated" dragon, we're hoping that it'll be more like a part of the family who actually wants to be well-behaved.

If you're thinking of making your own dragon, be sure that it feels like part of the family and less of a trophy, and that you develop a caring relationship with it. As a result, the dragon will rightly trust you and feel at home (hopefully, it then won't be as eager to kill you). If you plan a positive purpose for your dragon (a humanitarian effort maybe) then maybe they'll feel more useful and less like an oddball on display for the world to see.

What is the point of us even making a dragon? We discussed this in a more lighthearted way at the beginning of the book, in Chapter 1, but it warrants more thought here in this chapter on ethics. If our dragon is created as an intellectual exercise, is that ethical? Probably not. We don't imagine that our dragon would be a celebrity or that it would improve our own lives in a material sense, by earning lots of money for us, but once we've created the dragon, we may have less control over its life than we had hoped.

Would it be good for the dragons if we made them?

Some people think of human reproduction as an experiment of a sort and children are the results of that kind of experiment. In this way of thinking, children and, in fact, all of us humans and other animals, have no say in our own experimental creation. Who knows what combining different parental genomes will yield? There are also countless parental decisions, particularly by pregnant women, that greatly impact their future offspring. Other than extreme cases like doing drugs or drinking while pregnant, a surprisingly wide range of risks involved in reproduction aren't considered unethical.

Various kinds of plant and animal breeding to generate new species aren't unethical. While creating genetically modified organisms (GMOs) is more controversial, that too has become fairly widely accepted in some countries, even though again the new organisms cannot consent to their creation.

Along these lines of thinking, we believe it is not inherently unethical to bring dragons into existence. However, it's important to ask whether their creation is actually a good thing for the dragons themselves. What are the risks to them? Are there potential benefits to the dragons?

Risks of making malformed dragons along the way

One possible consequence of our dragon-making research, and particularly if it fails, is the creation of malformed or dead dragon-like creatures along the way.

Would it be ethical to try to create a dragon if we knew it was likely that our first efforts would result in malformed dragons? It's hard to say for certain, but we could do everything possible to reduce this risk and if it happens to limit how often it occurs moving forward.

One way to try to mitigate against such risks to the creatures we generate would be to carefully monitor their development in utero before they were born, for example, by ultrasound. If they appear to have serious developmental problems, we might have to terminate the pregnancy. But if we could not spot a problem before a creature was born, the most humane thing to do might be to euthanize them (as humanely as possible) once they were born.

We could, nevertheless, learn from such imperfect creatures as the research went forward, which would be crucial for making progress and for learning from our mistakes. It might also help to minimize the number of future imperfect animals that we produce.

Realistically, what are the odds that we'll get our dragon right on the first try? We think they are nearly zero. A big question then is: is it worse to kill a developmentally goofed up dragon before it's born to end possible suffering or to let it live to survive a possibly short, difficult or miserable life just so we can study it? Both options sound kind of awful and perhaps at least somewhat unethical in their own ways.

Admittedly, there is no real imperative to making a dragon. Some might even say it's essentially a vanity project that is of no benefit to humankind. On that basis, it arguably would be unethical to create damaged or dead dragon-like animals in pursuit of something that essentially offers no concrete benefit to other people or to the animals being used in the research.

Short-lived or sickly dragon?

What if our dragon seems fine at birth and even through its "childhood," but ends up living a tragically short life?

Or it lives a reasonably long time, but is regularly ill or chronically sick?

We certainly wouldn't want our dragon to be unwell. As we do our research and tinker with genetics and development, there are any number of ways in which our created dragon could end up being in poor health. For example, as we engineer its ability to breathe fire or to fly, it's possible that these additional traits could come at the expense of a robust immune system that can fight off infections.

Overall, there would be no guarantees in terms of how healthy or long-lived our dragon would turn out to be. It's going to take trial and error not just on the immune system front, but on every level. How well does its heart work? How do its lungs (throat, and nose) fare, particularly after all that fire-breathing? Does its gut handle all those flammable gases okay? How does its body handle flying?

These kinds of uncertainties together constitute another reason why it's likely that we'd need to make many dragons and then "fine tune" our methods from experience. Importantly, this will also require us to collect a large amount of data on our dragons in terms of their biological attributes and health. Will journals be willing to publish our data? It's hard to say. Will they or others raise ethical concerns over dragon research? We think this is likely, which will present further challenges. On the other hand, feedback from others will make the project better and likely more ethical.

Dragon blues?

And what if our dragon is profoundly unhappy or outright depressed or anxious?

There are many ways this could happen. If the dragon was stolen from us and used as the centerpiece of a zoo-like theme park, we can't imagine it being happy. Or our dragon might be stolen by governmental agencies, such as the US CIA (one of the main American spy agencies), or by militaries around the world. Their aim might be to use our dragon as a weapon in a way that our dragon wouldn't like. But to be honest, our dragon could end up unhappy with its life even if it wasn't stolen or had something else bad happen to it. This kind of thing happens to people too sometimes.

What then for our dragon? Therapy? Medication?

Would it be unethical if our dragon lived a long, but mostly unhappy, life?

A parent cannot guarantee the length of our children's lives or their happiness either. Sometimes, these things come down to chance and even to the decisions that an individual makes over the years. With our dragon though the situation would be a little different. We would have brought it into existence when it was previously not a real thing, and we'd have a huge input into the things that would have impact on its quality of life, like its genetic state and its environment.

Also, what if we end up making many dragons? Would some end up being sold to billionaires or to other countries? As we've said before, the dragons might be stolen too. What kind of life would await them? Governments or bad actors could use dragons in unethical ways such as for building armies. Both the dragon and many people could get hurt or killed.

Dragon benefits?

What specifically does our newly created dragon get out of this high-risk project?

At the most basic level, it gets to exist and experience the world. If it is fertile, it may become a parent and the founder of a future species on the planet. If our dragon is intelligent and not profoundly unhappy, it may enjoy its existence and find meaning in its life. Ideally, overall the dragon could be integrated into human society but at the same time have its own unique identity and place in the world. It wouldn't be attacked or forced to fight in wars.

How likely are these possible benefits and this potential overall rosy outcome? We cannot even begin to predict that until we actually do the project.

Benefits to the world?

We also have to ask ourselves, "Could making a dragon yield beneficial outcomes for the world?"

We could imagine our dragon having a positive impact on the world, for example, by inspiring far more young people in new generations to become scientists.

More practically speaking, the dragon might devote its life to benefiting others such as by flying medical supplies to those in need. But of course, you'd be right to ask, "why a dragon and not a plane?" We could respond with the idea that dragons are "a renewable resource."

Is this a possible way to justify the creation of these animals? Maybe. Realistically, a peaceful dragon whose life's mission is to benefit others might, be an oxymoron.

Is the possible "rebellion" of these dragons inevitable? It's not terribly unlikely.

Endangering the already at-risk Komodo?

Using animals in any type of research raises potential ethical issues. However, some animals require even more consideration. For instance, Komodo dragons are a somewhat endangered species and there aren't many of them roaming the planet. Would it be ethical to use a few endangered Komodos for our dragon-building experiments? It wouldn't be, would it? But what if we only used Komodos already in zoos or in private collections (does anyone seriously have a Komodo as a pet?) To us this latter option seems more permissible.

Instead of using real, intact Komodos, we also could just use their reproductive cells instead. Or cells from other parts of their body, obtained in a way that wouldn't cause a Komodo harm. For instance, a tiny skin sample could provide the cells needed to make those special, immortal stem cells called "IPS cells" (see Chapter 6). These IPS cells could then be used to make any other kind of cell, including sperm and eggs. Let's say we did this – used sperm and eggs created from IPS cells to generate a fertilized embryo. We'd still need a surrogate mother to carry the Komodo embryos in the process of making a dragon.

Would it be ethical to use a female Komodo for this purpose rather than letting her breed normally to make more Komodos given that they are endangered? If not, then we shouldn't use a Komodo as the surrogate mother. What then would we do? Use an alligator or crocodile? Try not to use a mother at all, and invent technologies to develop Komodo eggs entirely in the lab in a special incubator?

You can see how ethically complex this can get.

What if we could give millions of dollars to start or greatly expand a Komodo conservation program in the wild (significantly increasing their numbers) if we could, in exchange, use a few Komodos and/or their cells for our dragon research?

Fortunately, most of the other animals we have considered using as our "starter creatures" (such as various birds and Draco lizards) are not endangered, but we'd still have to do everything possible to maintain high ethical standards in how we use these animals and their cells.

Ideally, we'd want to have an ethics advisory board to help guide us. An advisory board of this kind would have members who are experts in bioethics and can provide us with informed advice about what we are doing or planning to do, or even tell us that we shouldn't do something we might be planning to do if they judged that what we were planning to do would be too dangerous, for example.

What if our dragons outlive us?

What if our dragon lives for hundreds of years – as was often the case in dragons of lore – and outlives us for many years? We would no longer be around to have a positive impact on its life nor influence what it does. After our deaths, our dragon might behave in ways that endanger our fellow humans – it might pillage and rampage once out of our control. Arguably, it would be unethical if we aren't around for our dragon. We would no longer be there to be responsible for and to care for or support a dangerous animal we had created.

It's also possible that our dragon could suffer and be very depressed and lonely after our deaths. However, in reality, the same kind of situation occurs for all of us who are parents who do not outlive our children. Our children must get along without us. In this spirit, we could make arrangements for our dragon's care as well as both financial and potentially emotional support should we both die before it does.

There are many ways in which trying to build a dragon raises ethical dilemmas. Even just the process of making the dragon itself is ethically dubious. But once the dragon is created and especially if it is still around after we are long gone, more challenges are likely to arise.

Ethics and government regulation

Governmental agencies play an important role in overseeing research, particularly certain research that produces genetically modified crops and animals. In the USA, the Food and Drug Administration (more commonly known as the FDA) plays a key role in regulating food, drugs and therapeutic devices in the interest of public health. They do so by overseeing and enforcing the many laws and regulations that control the production of food and drugs.

We believe it'd be important to follow the FDA's rules (if there are any relevant rules) when creating our dragon. However, sometimes governmental agencies fail to regulate the creation of new organisms by researchers. For instance, surprisingly, no governmental agency has – as yet – regulated the production of fluorescent, genetically modified GloFish, which we mentioned back in Chapter 5 (see Figure 5.5).

How is that possible?

In their article "CRISPR Critters and CRISPR Cracks," Hank Greely and Alta Charo explain how GloFish slipped (or rather swam) through the regulatory cracks [1]. Strangely, it turns out that the U.S. Environmental Protection Agency (EPA), the U.S. Department of Agriculture, and the U.S. Fish and Wildlife Service all said they had no jurisdiction over this new fish. And the FDA declined to review it, even though they could have regulated it as a "new animal drug" that should be tested for safety to itself and to the environment before being approved. Essentially, the FDA decided that GloFish were not going to be used for food, so no risk on that front, and were unlikely to be a threat to the environment.

What would various state and federal regulators think of our plans to create a dragon? And how would they view the dragon itself? Would the resulting dragon be a "product" (it's not a food or drug) that could be regulated by the FDA? Would any of the other regulatory agencies mentioned above be involved in regulating its creation and what it could subsequently do?

And what about agencies in other countries, such as the European Environmental Agency[i] and China's Ministry of Ecology and Environment, just to name two.[ii] We don't imagine our dragon or dragons will be limited by borders to just one country. It seems likely that some government agencies would frown on our efforts even if

---

[i] https://www.eea.europa.eu/

[ii] http://english.mee.gov.cn/

they don't have any specific rules or regulations that actually pertain to dragons.

Where do we get all the needed money without "selling out"?

Research projects, even those that are relatively simple, always cost far more than anyone predicts upfront and building a dragon is not going to be simple. While it's hard to pinpoint exact total costs, it is likely to be in the tens of millions of dollars for dragon production. At least a million just to start.

How do we get this funding without selling out to sketchy investors or governments that may then snatch our dragons or misuse them? If we are lucky, we might be able to find investors who are good people or companies that are focused on benefiting the world. In exchange, they might require us to make the dragons be peaceful or good citizens of the world, which isn't really a bad thing if we do want to be ethical about all of this. Still, it might be impossible to make our dragon behave in a specific way.

It's hard to imagine government funding agencies like the National Institutes of Health (NIH) or National Science Foundation (NSF) in the U.S. (or equivalent agencies abroad) being open to a grant application focused on building a dragon. While the American defense research agency, DARPA, might go for it, they would perhaps see the dragon as a potential weapon. We don't want that.

We suppose we could form a publicly-traded company to build a dragon and raise money by selling stock in our firm. However, the stock investors in our company (perhaps we'd call it Dragon X) might interfere with the project or try to get rid of our ethics advisory board. The stock investors might also end up wanting us to sell dragons to make a profit.

On the other hand, if we don't have the money up front, and go deeply in debt by building a dragon, we might be pressured to sell dragons as well. We also do not want to feel like we must have our dragons go on tour with a circus-like atmosphere to raise money either as that would negatively affect their quality of life.

Perhaps a responsible, fun, and educational dragon tour more akin to a traveling art exhibit could be enjoyable for the dragon and still raise a lot of money?

Clearly, the money side of dragon building is complicated and will require some more detailed financial planning.

Where we go, will others follow?

If we build a dragon and are transparent about the whole process, perhaps even publishing detailed methods about how to make a dragon in scientific journals (or if that is not possible, on the web), then could many other people follow in our footsteps, or at least try to? And if so, would we have opened a Pandora's box, popping out of which are not only dragons but also other wild new creatures like the unicorns mentioned in the previous chapter? Things could get very messy and interesting.

In some ways, it's not so different from building an atomic bomb for the first time. We can see from history that although the bomb was built in "secret," the details got out and others built atomic bombs too. The state of nuclear weapons technology and arms building across the globe today is not exactly a shining example of human technological wisdom.

And while we might say to ourselves "no, that's too much" in some cases, for example by not creating mythical creatures that are part human as mentioned in Chapter 7, others might not stop.

On the other hand, if *we* can think up how to make dragons and other mythological creatures like unicorns, then certainly others can too. Clearly, other people have already thought about the idea of making dragons even if they didn't write a whole book about it. Already, there is a DIY kind of movement based on CRISPR gene-editing that might have begun to introduce CRISPR gene-edited new creatures into the world.

A researcher in China, He Jiankui, went much further with CRISPR and claims to have created the first "gene-edited" human beings. He has said that he introduced CRISPR into a variety of human embryos and reportedly (none of this has been independently confirmed) ended up with two girls with somewhat random chunks missing in a certain gene called *CCR5* that impacts how susceptible people are to HIV infection.

To be clear his effort was misguided and poorly planned. We think it was unethical too. There are already proven, safe ways to prevent transmission of HIV, and He Jiankui's twins may not actually be resistant to HIV given the randomness of the genetic changes that were made. They are also at risk for other health problems. He himself is being investigated by Chinese authorities and is likely in serious trouble. He might be prohibited from doing research or could end up in jail. You can read more about He Jiankui's rogue CRISPR of humans here on Paul's blog The Niche.[iii]

Think CRISPR will be limited just to highly-regulated research labs? That's not the reality. You can even already order a "CRISPR kit" online to DIY gene edit organisms.[iv] While at the moment, this particular kit is limited for use just on microorganisms, it could lead to the genetic

---

[iii] https://ipscell.com/?s=he+jiankui

[iv] https://www.scientificamerican.com/article/mail-order-crispr-kits-allow-absolutely-anyone-to-hack-dna/

modification of more complex animals in the near future. Others are gearing up to sell similar kits that might be used in more outlandish ways.

Perhaps the Pandora's box of building new creatures is already open due to technological advances. In that sense, this book is also a wake-up call. While people aren't building dragons (that we know of), there are numerous people trying to make all kinds of new organisms or make changes in themselves.

While this chapter has made it clear that there would be numerous ethical challenges (some of which are almost unresolvable) associated with trying to build a real dragon, again the point of this book is not to actually build or help other people make real dragons. Instead, our goal has been to have fun with readers in new areas of science, poke satirical fun at how much science is hyped, and encourage people around the world to use their imaginations with science, while keeping bioethics in mind too.

Reference

1.　　Charo, R.A. and H.T. Greely, CRISPR Critters and CRISPR Cracks. *Am J Bioeth* 2015. **15**(12): 11–17.

Glossary

Archaeopteryx. A genus of winged, bird-like dinosaurs. Some see them as an evolutionary link between non-winged dinosaurs and modern birds.

Bombardier beetle. A unique type of insect that can shoot hot chemical liquid out of its rear end as a defense mechanism, which can burn predators.

Brown fat. A rare common type of fat (as compared to the general "white fat") that can be used by the body to generate body heat. It is far more common in babies than in adults, some of whom may have no brown fat at all.

Cerebellum. The back region of the brain that is important for hand-eye coordination and proper movement. The word literally means "little brain."

Chimera. An animal composed of parts of two or more different types of animals. With only rare exceptions, chimeras are mythological creatures. In theory, an animal entirely composed of its own cells, but containing inserted genes of other species could be considered a "genetic chimera."

Cloning. The process of making a duplicate, entirely new organism from a single cell taken from an adult animal. Cloning has been achieved in frogs, farm animals, and some other species studied in the lab such as mice. To our knowledge, human cloning has not been accomplished. "Cloning" can also refer to the production of embryonic stem cells containing the nuclei of a different person.

Conjoined twins. A unique type of twins where the two twins' bodies are at least partially fused together, and the twins share some body parts. This pregnancy outcome results from problems during very early embryo development.

CRISPR gene-editing. A system used to make genetic modifications or mutations in specific genes in cells or entire organisms. It is sometimes called "CRISPR-Cas9 gene-editing."

Draco lizards. A small type of lizard with skin flaps (patagia) that allow it to soar from tree to tree like a flying squirrel.

Electrocytes. Special cells present in animals with electrical organs in their bodies, such as electric eels, that generate electricity that can then be used by the animal to shock prey or sense surroundings.

Embryonic stem cells. These cells, often called "ES cells," in the human case are derived from human embryos that are leftover in freezers from fertility procedures after couples undergoing IVF have already successfully had children. They are thought to have the ability to be made into any type of cell in the body. Embryonic stem cells have not successfully been made from very many types of animals, but in principle it is thought that many animals could have their own version of ES cells produced.

Eyeshine. The quality of some animals' eyes whereby they shine at night when light is directed towards them due to a specific eye structure called the tapetum lucidum. Animals with eyeshine have superior night vision compared to those that don't such as people.

Gastroliths. Small stones that are swallowed by some animals and that can aid in digestion.

Gizzard. A special part of the stomach in some animals, particularly birds but also others including potentially dinosaurs, that is or was mainly used to grind foods. The gizzard can sometimes contain gastroliths to aid in grinding.

Hatzegopteryx. A very large Pteranodon that was a bit smaller than its relative Quetzalcoatlus.

Induced pluripotent stem cells. These cells, often going by the nickname "iPS cells," are just like embryonic stem cells in terms of their power to make other cell types, but they are not made from embryos. Instead, they are produced from ordinary non-stem cells through a process called "reprogramming."

IVF or *in vitro* fertilization. A method to fertilize eggs with sperm to generate embryos outside the body, which in humans can then be implanted into a surrogate mother to produce a baby.

Keratin. A major structural protein present in skin (and related structures including hair, nails, feathers, and scales) and in specialized cells involved in growth and the function of these tissues.

Keratinocytes. Specialized skin and other cells that have a high abundance of keratin.

Komodo dragon. An extremely large and sometimes dangerous type of lizard found in Indonesia.

Melanin. The main pigment in humans and many other animals that gives each individual the unique colors of their skin, hair, and other tissues. Albinos lack melanin.

Melanocytes. The cells that make melanin.

Microbiome. A collection of microbes such as bacteria within a larger organism such as in the guts of humans or other animals, or more specifically the genomes of these microbes.

Microcephaly. A condition where the brain and head do not grow normally and end up too small. It is often, but not always, associated with impaired intelligence. Causes include infections such as by the Zika virus after mosquito bites to pregnant women in some parts of the world and also genetic mutations.

Nictitating membrane. A unique transparent extra eyelid in some animals that serves a protective function for the eye.

Organoids. Miniature organs that can be grown in the lab from stem cells. For instance, human brain organoids are miniature versions of the developing human brain that can be made from human stem cells.

Parthenogenesis. The development of a new animal independent of fertilization by sperm, which can happen if an unfertilized egg begins dividing. While this process occurs at times in some species, it has never been reported in humans.

Patagium (plural: patagia). A large flap of skin, often between fingers but also between arms or legs and the body, which aids in flight.

Pteranodon. A genus of extremely large flying pterosaurs.

Quetzalcoatlus. An extinct Pteranodon that may have been the largest flying creature in Earth's history.

Regeneration. The ability for an animal to regrow part of its body or a certain tissue that has been lost.

Rumen. The top part of the stomach in some animals such as in cattle that starts digestion.

Stem cells. Relatively rare cells in the body that can either make more of themselves by cell division or change into other cell types such as neurons, muscle, etc. through a process called differentiation.

Syndactyly. A condition where individual digits such as fingers or toes in humans are fused or connected by webbing.

***Wnt* genes.** Genes that in winged animals are involved in wing growth and development, but that in non-winged animals including humans play other roles in tissue development.

Index

CPSIA information can be obtained
at www.ICGtesting.com
Printed in the USA
LVHW080631091219
639848LV00006B/123/P

9 789813 275935